Current Topics in
Developmental Biology
Volume 50

Series Editor

Gerald P. Schatten
Departments of Obstetrics–Gynecology
and Cell and Developmental Biology
Oregon Regional Primate Research Center
Oregon Health Sciences University
Beaverton, Oregon 97006-3499

Editorial Board

Peter Grüss
Max-Planck-Institute of Biophysical Chemistry
Göttingen, Germany

Philip Ingham
University of Sheffield, United Kingdom

Mary Lou King
University of Miami, Florida

Story C. Landis
National Institutes of Health/
National Institute of Neurological Disorders and Stroke
Bethesda, Maryland

David R. McClay
Duke University, Durham, North Carolina

Yoshitaka Nagahama
National Institute for Basic Biology, Okazaki, Japan

Susan Strome
Indiana University, Bloomington, Indiana

Virginia Walbot
Stanford University, Palo Alto, California

Founding Editors

**A. A. Moscona
Alberto Monroy**

Current Topics in Developmental Biology

Volume 50

Edited by

Gerald P. Schatten

*Departments of Obstetrics–Gynecology
and Cell and Developmental Biology
Oregon Regional Primate Research Center
Oregon Health Sciences University
Beaverton, Oregon*

ACADEMIC PRESS

A Harcourt Science and Technology Company

San Diego   San Francisco   New York   Boston   London   Sydney   Tokyo

Cover photo credit: GUS activity in shoot apex of a two-day-old seedling. From Figure 6 in Chapter 3 "Mechanisms of Plant Embryo Development" by Shunong Bai, Lingjing Chen, Mary Alice Yund, and Zinmay Renee Sung. See text and color plate for further details.

This book is printed on acid-free paper.

Copyright © 2000 by ACADEMIC PRESS

All Rights Reserved.
No part of this publication may be reproduced or transmitted in any form or by any means, electronic or mechanical, including photocopy, recording, or any information storage and retrieval system, without permission in writing from the Publisher.

The appearance of the code at the bottom of the first page of a chapter in this book indicates the Publisher's consent that copies of the chapter may be made for personal or internal use of specific clients. This consent is given on the condition, however, that the copier pay the stated per copy fee through the Copyright Clearance Center, Inc. (222 Rosewood Drive, Danvers, Massachusetts 01923), for copying beyond that permitted by Sections 107 or 108 of the U.S. Copyright Law. This consent does not extend to other kinds of copying, such as copying for general distribution, for advertising or promotional purposes, for creating new collective works, or for resale. Copy fees for pre-2000 chapters are as shown on the title pages. If no fee code appears on the title page, the copy fee is the same as for current chapters.
0070-2153/00 $35.00

Explicit permission from Academic Press is not required to reproduce a maximum of two figures or tables from an Academic Press chapter in another scientific or research publication provided that the material has not been credited to another source and that full credit to the Academic Press chapter is given.

Academic Press
A Harcourt Science and Technology Company
525 B Street, Suite 1900, San Diego, California 92101-4495, USA
http://www.academicpress.com

Academic Press
Harcourt Place, 32 Jamestown Road, London NW1 7BY, UK
http://www.academicpress.com

International Standard Book Number: 0-12-153150-3

PRINTED IN THE UNITED STATES OF AMERICA
00 01 02 03 04 05 EB 9 8 7 6 5 4 3 2 1

Contents

Contributors   ix
Preface   xi

1
Patterning the Early Sea Urchin Embryo
Charles A. Ettensohn and Hyla C. Sweet

I. Introduction   1
II. An Overview of Patterning in the Sea Urchin Embryo   2
III. Fate Maps of the Sea Urchin Embryo   3
IV. Gene Expression Maps of the Early Sea Urchin Embryo   7
V. Mechanisms of Patterning along the Animal–Vegetal Axis   10
VI. Mechanisms of Patterning along the Oral–Aboral Axis   24
VII. A Model of Early Patterning   32
    References   36

2
Turning Mesoderm into Blood: The Formation of Hematopoietic Stem Cells during Embryogenesis
Alan J. Davidson and Leonard I. Zon

I. Introduction   45
II. Mesoderm Induction and Patterning   46
III. Fate Maps   48
IV. Ventralizing Homeobox Genes   49
V. Specification of Hematopoietic Stem Cells   52
VI. Conclusions   53
VII. Future Directions   55
    References   55

3
Mechanisms of Plant Embryo Development
Shunong Bai, Lingjing Chen, Mary Alice Yund, and Zinmay Renee Sung

I. Introduction   61
II. Embryogenesis   62

III. Histogenesis: The Generation of Cells and Tissues   65
IV. Organogenesis: The Generation of Organs   71
V. Embryo-Specific Genes and Embryonic Mutants   76
VI. Embryonic Induction   78
VII. Summary   83
References   83

4
Sperm-Mediated Gene Transfer
Anthony W. S. Chan, C. Marc Luetjens, and Gerald P. Schatten

I. Introduction   89
II. DNA Binding   90
III. DNA Internalization   91
IV. DNA Integration   92
V. Alternative Strategies   94
VI. Perspectives   98
References   99

5
Gonocyte–Sertoli Cell Interactions during Development of the Neonatal Rodent Testis
Joanne M. Orth, William F. Jester, Ling-Hong Li, and Andrew L. Laslett

I. The Perinatal Period of Testicular Development: Historical Perspective   104
II. The Sertoli Cell–Gonocyte Coculture Model   106
III. Cell–Cell Interactions in Neonatal Gonocyte Development   110
IV. Summary   122
References   122

6
Attributes and Dynamics of the Endoplasmic Reticulum in Mammalian Eggs
Douglas Kline

I. The Endoplasmic Reticulum (ER), Calcium, and Egg Activation   126
II. Development of the Calcium-Releasing System during Oocyte Maturation   131
III. Arrangement and Reorganization of the ER   133
References   147

7
Germ Plasm and Molecular Determinants of Germ Cell Fate
Douglas W. Houston and Mary Lou King

I. Introduction   156
II. Germ Plasm   157

Contents

  III. Experimental Evidence for the Role of Germ Plasm in Primordial
       Germ Cell Formation   163
  IV. Molecules Localized to the Germ Plasm   165
   V. Summary and Future Directions   174
      References   176

Index   183
Contents of Previous Volumes   191

Contributors

Numbers in parentheses indicate the pages on which the authors' contributions begin.

Shunong Bai,[1] Department of Plant and Microbial Biology, University of California at Berkeley, Berkeley, California 94720 (61)

Anthony W. S. Chan, Oregon Regional Primate Research Center, Beaverton, Oregon 97006 (89)

Lingjing Chen, Department of Plant and Microbial Biology, University of California at Berkeley, Berkeley, California 94720 (61)

Alan James Davidson, Division of Hematology/Oncology, Children's Hospital, Boston, Massachusetts 02115 (45)

Charles A. Ettensohn, Department of Biological Sciences, Carnegie Mellon University, Pittsburgh, Pennsylvania 15213 (1)

Douglas W. Houston,[2] Department of Cell Biology and Anatomy, University of Miami School of Medicine, Miami, Florida 33101 (155)

William F. Jester, Department of Anatomy and Cell Biology, Temple University School of Medicine, Philadelphia, Pennsylvania 10140 (103)

Mary Lou King, Department of Cell Biology and Anatomy, University of Miami School of Medicine, Miami, Florida 33101 (155)

Douglas Kline, Department of Biological Sciences, Kent State University, Kent, Ohio 44242 (125)

Andrew L. Laslett, Department of Anatomy and Cell Biology, Temple University School of Medicine, Philadelphia, Pennsylvania 10140 (103)

Ling-Hong Li,[3] Department of Anatomy and Cell Biology, Temple University School of Medicine, Philadelphia, Pennsylvania 10140 (103)

C. Marc Luetjens, Center for Interdisciplinary Clinical Research, Research Group "Programmed Cell Death," Westphalian Wilhelms University, D-48148 Münster, Germany (89)

[1]Present address: Department of Plant Molecular and Developmental Biology, Beijing University, Beijing 100871, People's Republic of China.
[2]Present address: Division of Developmental Biology, Children's Hospital Research Foundation Cincinnati, Ohio 45229.
[3]Present address: OEHHA/California EPA, Oakland, California 94588.

Joanne M. Orth, Department of Anatomy and Cell Biology, Temple University School of Medicine, Philadelphia, Pennsylvania 10140 (103)

Gerald P. Schatten, Oregon Regional Primate Research Center, Beaverton, Oregon 97006 (89)

Zinmay R. Sung, Department of Plant and Microbial Biology, University of California at Berkeley, Berkeley, California 94720 (61)

Hyla C. Sweet, Department of Biological Sciences, Carnegie Mellon University, Pittsburgh, Pennsylvania 15213 (1)

Mary Alice Yund, Department of Plant and Microbial Biology, University of California at Berkeley, Berkeley, California 94720 (61)

Leonard Ira Zon, Division of Hematology/Oncology, Children's Hospital, Boston, Massachusetts 02115 (45)

Preface

Current Topics in Developmental Biology is turning 50! Volume 50 is presented as a milestone marking the continuation of providing comprehensive surveys of major issues, while also staying at the forefront of modern developmental biology. Volume 50 also marks a transition in the editorial leadership. Professor Roger Pedersen of the University of California, San Francisco, who has served as the sole editor, and later the co-editor, for the past decade, has chosen to relinquish this post. The field of developmental biology owes Roger a great debt of gratitude for his dedication, enthusiasm, insights, and wisdom in guiding CTDB, and we will continue to rely on Roger for thoughtful and sound advice.

This 50th volume continues the tradition of this series in addressing developmental mechanisms in a variety of experimental systems. The conceptual sequence of topics begins with totipotency by examining germ plasm, germ cell fates, and gonocytes, continues with sperm-mediated transgenic approaches and endomembrane dynamics within eggs, and then considers embryonic patterning in both plants and animals as well as pluripotency of hematopoietic stem cells. The chapter by Charles A. Ettensohn and Hyla C. Sweet reviews current knowledge of the molecular mechanisms of embryonic patterning in sea urchins and provides molecular mechanisms for animal–vegetal as well as oral–arboral axes. Alan J. Davidson and Leonard I. Zon examine the development biology of "Turning Mesoderm into Blood: The Formation of Hematopoietic Stem Cells during Embryogenesis." The implications of developmental medicine are truly revolutionary.

"Mechanisms of Plant Embryo Development," by Shunong Bai, Lingjing Chen, Mary Alice Yund, and Zinmay Renee Sung, integrates classic studies on morphogenesis with current molecular concepts. Anthony W. S. Chan, C. Marc Luetjens, and I consider the role of sperm as a vector for transgenesis. Joanne M. Orth, William F. Jester, Ling-Hong Li, and Andrew L. Laslett present "Gonocyte–Sertoli Cell Interactions during Development of the Neonatal Rodent Testis." Douglas Kline considers the attributes and dynamics of the endoplasmic reticulum in mammalian eggs. Finally, Douglas W. Houston and Mary Lou King present "Germ Plasm and Molecular Determinants of Germ Cell Fate."

These chapters should be valuable to researchers in the fields of animal and plant development, as well as to students and other professionals who want an introduction to current topics in cellular and molecular approaches to developmental biology. This volume in particular will be essential reading for anyone interested in stem cells—including hematopoietic and germ cells—transgenic approaches, embryonic patterning in animals and in plants, and membrane dynamics during development.

This volume has benefited from the ongoing cooperation of a team of participants who are jointly responsible for the content and quality of its material. The authors deserve the full credit for their success in covering their subjects in depth yet with clarity and for challenging the reader to think about these topics in new ways. We thank the members of the Editorial Board for their suggestions of topics and authors.

Gerald P. Schatten

1
Patterning the Early Sea Urchin Embryo

Charles A. Ettensohn and Hyla C. Sweet
Department of Biological Sciences and
Science and Technology Center for Light Microscope Imaging and Biotechnology
Carnegie Mellon University
Pittsburgh, Pennsylvania 15213

I. Introduction
II. An Overview of Patterning in the Sea Urchin Embryo
III. Fate Maps of the Sea Urchin Embryo
IV. Gene Expression Maps of the Early Sea Urchin Embryo
V. Mechanisms of Patterning along the Animal–Vegetal Axis
   A. Maternal Patterning
   B. Cell–Cell Interactions
VI. Mechanisms of Patterning along the Oral–Aboral Axis
   A. Expressions of the Oral–Aboral Axis
   B. Blastomere Isolation/Recombination Studies
   C. BMP2/4 Signaling
   D. Analysis of *cis*-Regulatory Regions of Genes Differentially Expressed along the Oral-Aboral Axis
VII. A Model of Early Patterning
   References

I. Introduction

This review summarizes our current understanding of mechanisms that underlie the patterning of the early sea urchin embryo. Its focus is on the partitioning of the cleavage and blastula stage embryo into distinct domains of gene expression and cell fate. Although many questions remain unanswered, recent studies have advanced dramatically our understanding of the early patterning of this embryo, and so a reevaluation of the problem is warranted.

Many of the terms related to cell fate specification and patterning have been assigned various meanings. *Allocation* will be used to refer to a clonal restriction in cell fate, as determined by fate mapping studies (Slack, 1991). A *founder cell* is the mother cell of a lineage allocated to a specific fate or a particular developmental state (e.g., the founder cell of an early embryonic territory as defined by the expression of a particular marker gene). Neither term describes the state of commitment of cells, which must be tested experimentally. *Determination* refers to the stable commitment of a cell to a particular fate, as assayed by cell isolation and cell transplantation experiments. *Specification*, which has been used variously by different authors (e.g., Davidson, 1989; Slack, 1991), will be used in a general sense to mean the progressive restriction of developmental potential.

The sea urchin embryo has a long and rich history as a model system for the analysis of patterning. This discussion focuses on early patterning events, i.e., those that take place prior to the start of gastrulation, and on recent studies. A number of reviews consider the older literature and/or selected aspects of patterning, including cell interactions that take place after the start of gastrulation, which are beyond the scope of this review (Hörstadius, 1973; Davidson, 1989; Ettensohn, 1992; Ettensohn *et al.*, 1997; Henry, 1998). Current views of early fate specification have also been presented by Logan and McClay (1998), Davidson *et al.* (1998), and Angerer and Angerer (2000).

II. An Overview of Patterning in the Sea Urchin Embryo

Echinoderms, which include the sea urchins, are deuterostomes and the closest living relatives of the chordates. Most sea urchin species commonly used for developmental studies exhibit indirect development, a mode that leads to the formation of the adult via a feeding pluteus larva. The free-swimming larva consists of approximately 12 different cell types organized in a stereotypical body plan (Fig. 1). The larva has three axes of asymmetry: the anterior–posterior axis (roughly aligned with the original animal–vegetal axis of the egg), the oral–aboral (or ventral–dorsal) axis, and the right–left axis.

Experimental studies have focused primarily on the embryonic phase of development, which culminates in the formation of the pluteus. Development is far from complete, however, at this stage. As the larva feeds and grows, it elaborates new anatomical structures, including the echinus rudiment, which will give rise to much of the adult sea urchin. Eventually, the larva undergoes metamorphosis and the radially symmetrical juvenile emerges from the remnants of the larval body. Relatively little is known about the complex developmental program that transforms the larva into the adult, although there have been morphological descriptions of the process (see Burke, 1989; Davidson *et al.*, 1998) and initial efforts are being made to analyze the expression of potential regulatory molecules during this phase of development (Davidson *et al.*, 1998).

Direct development, which transforms the egg into a juvenile sea urchin without an intervening larval stage, has arisen independently in at least six lineages of sea urchins (Raff, 1992). Even very closely related species can differ with respect to whether they exhibit direct or indirect development. Analysis of the cellular and molecular mechanisms of these alternative modes is providing insights into the way in which developmental programs have been modified during evolution (Raff, 1992). For example, there are several significant differences between indirect and direct developing embryos with respect to egg size, the fates of early blastomeres, axis determination, and patterns of gene expression (Henry and Raff,

1990; Wray and Raff, 1990; Raff, 1992; Klueg *et al.*, 1997; Kissinger and Raff, 1998; Raff *et al.*, 1999). Primarily because of these substantive differences, this review focuses on the more intensively studied, indirect developing species.

III. Fate Maps of the Sea Urchin Embryo

The classical fate map of the cleavage stage embryo (Hörstadius, 1973) has been modified in important ways by recent studies. Improved methods of cell labeling have been used to generate higher resolution fate maps and have made it possible to examine more advanced developmental stages. Furthermore, fate mapping studies have now been carried out in several species, allowing conclusions to be drawn about those features that are shared and those that are variable. The most extensive fate mapping studies have been carried out with embryos of *Lytechinus variegatus* and *Strongylocentrotus purpuratus*. Only the former will be described in detail here, with comparisons to *S. purpuratus* and other species.

As in other euechinoid species, early cleavage divisions in *L. variegatus* are stereotypical and give rise to distinct tiers of blastomeres along the animal–vegetal (A-V) axis (Fig. 2). This "textbook" view obscures the subtle variability of certain aspects of cleavage. For example, in some species, the position of the third cleavage plane can be variably equatorial or subequatorial. In *L. variegatus*, the tiers of cells that form in the animal region of the embryo are not arranged as regularly as those in the vegetal region and the an1/an2 layers can be difficult to identify reliably (Summers *et al.*, 1993).

Fate maps of the *L. variegatus* embryo are shown in Fig. 2. As determined by classical studies, the animal region of the embryo gives rise to the ectoderm and the vegetal region primarily to the mesoderm and endoderm. It was previously thought that the endoderm and mesoderm of the embryo are derived exclusively from the micromeres and the veg2 tier (Hörstadius, 1973). Fate mapping studies by Logan and McClay (1997), however, have shown that veg1 cells contribute not only to the ectoderm, but also to the endoderm; i.e., the ectoderm/endoderm boundary lies within the veg1 territory. The contribution of veg1 cells to the endoderm has been confirmed in *S. purpuratus* (Ransick and Davidson, 1998). In both *S. purpuratus* and *L. variegatus*, veg1-derived cells that will give rise to the endoderm are recruited into the archenteron beginning at the midgastrula stage, as part of the limited involution that takes place in the sea urchin (reviewed in Ettensohn, 1999; see Fig. 2). In *L. variegatus*, involution is radially symmetrical and the archenteron is centered in the blastocoel. In *S. purpuratus*, however, the number of involuting cells and, consequently, the contribution of veg1-derived cells to the archenteron is greater on the oral side of the embryo (Ransick and Davidson, 1998). This asymmetric involution may contribute to the displacement of the archenteron toward the oral surface in this species.

The revised fate map places the ectoderm–endoderm boundary closer to the

Text continues on page 6

Fig. 1 The principal cell types of the sea urchin embryo, illustrated in a lateral, cutaway view of an early pluteus larva. The three axes of polarity of the larva are shown: the oral–aboral (O-Ab) axis, the left–right (L-R) axis, and the original animal–vegetal axis of the early embryo (dotted line), which corresponds roughly to the anterior–posterior axis of the larva.

Fig. 2 Fate maps of the sea urchin embryo (*Lytechinus variegatus*). Prospective ectoderm, mesoderm, and endoderm are shown in shades of blue, red, and yellow, respectively. Green represents the prospective ciliated band. A, animal pole; V, vegetal pole; O, oral pole; Ab, aboral pole. (A) Fate map of the 60-cell stage embryo. The ectoderm (dark blue, oral ectoderm; light blue, aboral ectoderm) derives from the an1, an2, and veg1 layers. The prospective endoderm (yellow) is located subequatorially and is derived from both the veg1 and veg2 layers. Veg2 contributes primarily to the foregut and midgut, whereas veg1 contributes primarily to the midgut and hindgut (Logan and McClay, 1997). The mesoderm forms from the vegetal-most cells of the early embryo. The nonskeletogenic mesoderm (pink) is derived almost exclusively from the veg2 layer, with the exception of a small contribution by the small micromeres to the coelomic pouches (dark red) (Pehrson and Cohen, 1986). The large micromeres (red) give rise exclusively to skeletogenic (primary) mesenchyme cells (PMCs). (B) Fate maps of the vegetal region of the embryo, based largely on Ruffins and Ettensohn (1996) and Logan and McClay (1997). Lineage compartments are indicated by solid black lines. The left diagram shows the distribution of cell fates within the vegetal plate of the mesenchyme blastula stage embryo. At the center of the vegetal plate are the eight small micromere descendants (sm), which are surrounded by veg2-derived, prospective nonskeletogenic mesoderm. (Note that the skeletogenic primary mesenchyme cells have ingressed by this stage are not shown in this surface view.) Fates within the mesodermal region are polarized along the O-Ab axis. Prospective pigment cells are concentrated in the aboral region of the vegetal plate (stippled pink), whereas prospective blastocoelar cells are concentrated in the oral region (solid pink). The initial invagination of the vegetal plate involves the inpocketing of the prospective nonskeletogenic mesoderm, as shown by the position of the blastopore margin at the very early gastrula stage (dotted line). The prospective foregut, midgut, and hindgut endoderm (yellow, orange, and tan, respectively) are positioned as expanding concentric rings around the prospective mesoderm at the mesenchyme blastula stage. The middle and right diagrams show the progressive distortion of the fate map associated with the involution of cells during gastrulation. The solid black circle represents cells that have moved into the interior of the embryo during gastrulation. (C) Cell fates at the pluteus larva stage. This figure is similar to Fig. 1, but has been colored to facilitate correlation with the fate maps in A and B. Note that the coloration of the coelomic pouches is intended solely to indicate their dual origin from small micromeres and veg2 cells, and does not reflect the actual positions of cells in the rudiment.

Fig. 3 Gene expression maps of the egg and early embryo. (A) Unfertilized egg, showing maternal mRNAs known to have asymmetric patterns of expression. The *bep1*, *bep3*, and *bep4* mRNAs are localized in the animal region of the embryo (tan region), whereas the *COLL1α* and *Sp-COUP* mRNAs both show polarized distributions orthogonal to the A-V axis (red hatching). For simplicity, the expression patterns of *COLL1α* and *Sp-COUP* are shown as coincident, but this has not been demonstrated. (B) Morula (~128-cell stage). Restricted patterns of expression of early, zygotically activated genes are evident by this stage. These include the *VEB* genes, which are expressed in the animal two-thirds to three-fourths of the embryo (green region) (Lepage *et al.*, 1992a,b; Reynolds *et al.*, 1992; Angerer and Angerer, 1997). The vegetal boundary of *VEB* gene expression has not been precisely mapped with respect to known lineage boundaries, but appears to lie within the veg1 or veg2 compartments. The *PlHbox12* gene is expressed orthogonal to the A-V axis (purple hatching) (DiBernardo *et al.*, 1995). Its pattern of expression with respect to the O-Ab axis and to maternal mRNAs such as *Sp-COUP* or *COLL1α* has not been determined. (C) Mesenchyme blastula. The complexity of the gene expression map has increased markedly by this stage. There is evidence for the regional expression of several aboral ectoderm-specific mRNAs (blue region), including the CyIIIa/CyIIIb and Spec1/Spec2

Figure 3 legend continues on page 5.

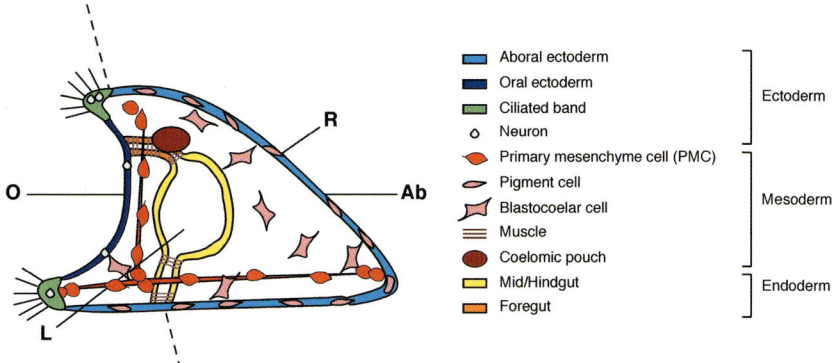

Chapter 1, Figure 1

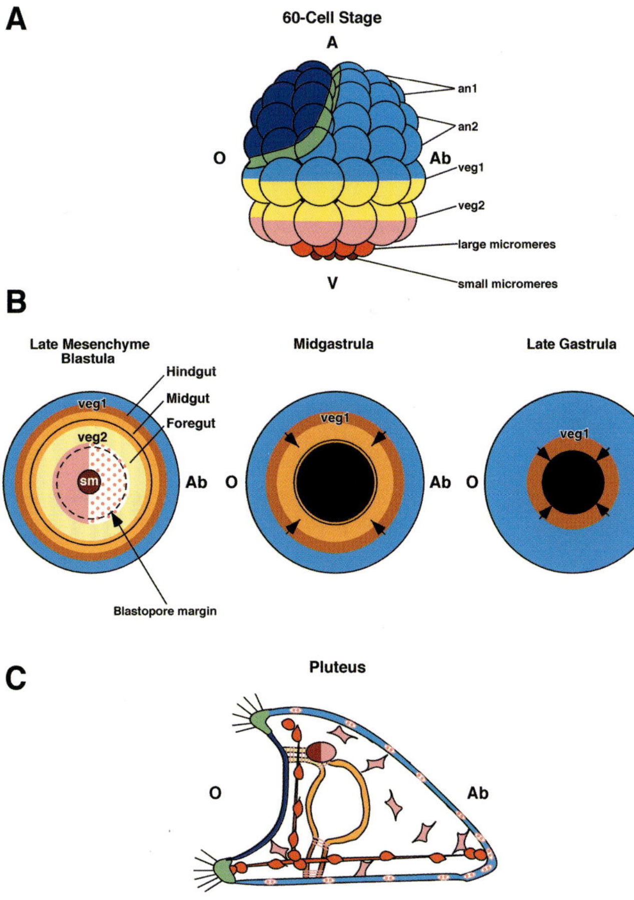

Chapter 1, Figure 2

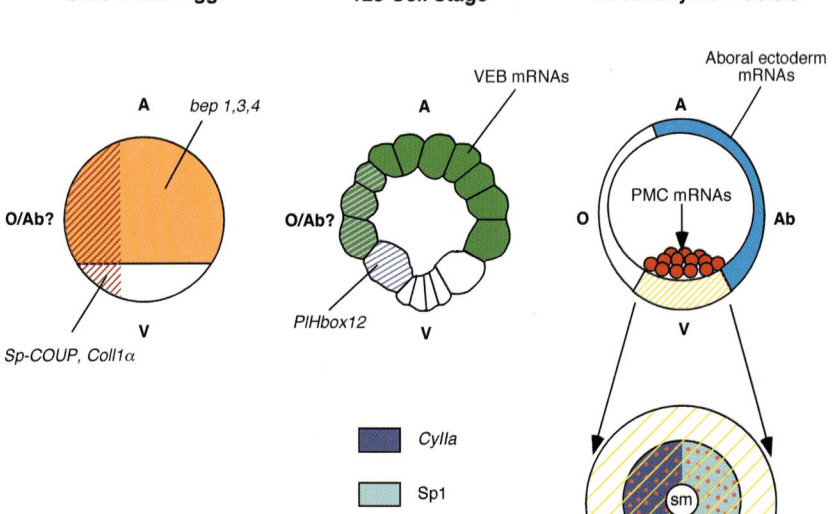

Chapter 1, Figure 3

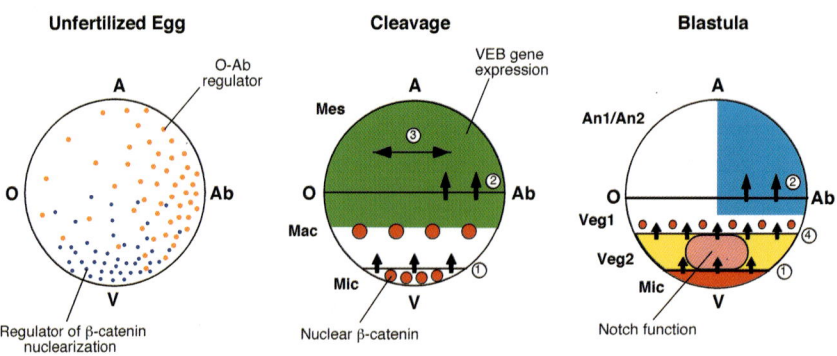

Chapter 1, Figure 4

Figure 3 continued
gene groups and arylsulfatase (reviewed by Angerer *et al.*, 1990). The skeletogenic PMCs (red) are expressing high levels of several cell-type-specific mRNAs, including *SM50* (Benson *et al.*, 1987), *PM27* (Harkey *et al.*, 1995), and *msp130* (Harkey *et al.*, 1992). The vegetal plate (yellow hatching) shows a complex pattern of gene expression (bottom enlargement). Several genes, such as *Endo16, S9, Brachyury,* and *SpKrox*, are expressed broadly throughout most of the vegetal plate (yellow hatching) (Ransick *et al.*, 1993; Harada *et al.*, 1995, 1996; Miller *et al.*, 1996; Wang *et al.*, 1996). Other markers are restricted in their expression to the prospective nonskeletogenic mesoderm (orange stippling). The SMC1 antigen is expressed specifically within this region, coincident with a domain of Notch protein downregulation (Sherwood and McClay, 1997; Sweet *et al.*, 1999). Within the prospective nonskeletogenic mesoderm, *CyIIa* transcripts (purple) accumulate on the oral side of the vegetal plate (Miller *et al.*, 1996). The Sp1 antigen (light blue) is restricted to prospective pigment cells (Gibson and Burke, 1985), which are concentrated aborally (Ruffins and Ettensohn, 1996). The small micromere (sm) territory (white region at the vegetal pole) is distinguished primarily by a lack of expression of other vegetal plate markers (Ransick *et al.*, 1993; Miller *et al.*, 1996, Sweet *et al.*, 1999), although these cells have been shown to have high levels of nuclear β-catenin (Miller and McClay, 1997).

Fig. 4 A model for the early patterning of the sea urchin embryo. Horizontal black lines indicate lineage compartments. Arrows show cell interactions, and colored regions indicate domains of gene expression. (Left) The unfertilized egg. Maternal patterning information along the A-V axis is proposed to exist in the form of a differential localization of a regulator of β-catenin nuclearization, e.g., a vegetally concentrated inhibitor of GSK3 activity (blue circles). A regulator of O-Ab axis formation (orange circles) shows a polarized distribution orthogonal to the A-V axis. (Middle) Cleavage. The activity of the maternal regulator leads to the accumulation of β-catenin in the nuclei of macromeres and micromeres (red circles). Levels of β-catenin control the polarized expression of *VEB* genes (green) along the A-V axis by a cell-autonomous mechanism. In the micromeres, nuclear β-catenin is required for the activation of the skeletogenic program of gene expression and for the signaling properties of the cells. We propose that in the macromeres and their progeny, nuclear β-catenin functions to (1) entrain a basal, micromere-independent program of endomesoderm formation, (2) endow the cells with signaling properties, and (3) increase their responsiveness to micromere-derived inductive signals, which are overlaid onto the basal program of endomesoderm formation. Micromere-derived signals (interaction No. 1) are required for the normal formation of both mesoderm and endoderm. Signals from the vegetal hemisphere (interaction No. 2) are required to polarize the overlying ectoderm along the O-Ab axis. We propose that these signals are themselves polarized and are dependent on the nuclearization of β-catenin in the vegetal region of the embryo for their establishment. Homotypic interactions among mesomeres (interaction No. 3) suppress their endomesodermal-forming potential. (Right) Blastula. Signals from the micromeres to the veg2 layer (interaction No. 1) are required for the down regulation of Notch protein in the central region of the vegetal plate, which, in turn, is required for the formation of the nonskeletogenic mesoderm. In contrast, the induction of endoderm by micromeres appears to occur by a Notch-independent mechanism. Concurrent with Notch downregulation, β-catenin accumulates in the nuclei of veg1-derived prospective endoderm (red circles), probably as a consequence of veg2-to-veg1 signaling (interaction No. 4). (Note that high levels of nuclear β-catenin also persist in the small micromeres, but this is not shown.) Interactions between vegetal and animal hemispheres lead to the polarized expression of aboral ectoderm-specific transcripts (blue territory). No oral ectoderm markers have been shown to be regionally expressed as early as the blastula stage.

animal pole than was previously predicted and shows that the allocation of cells to these two germ layers is not tightly linked to early cleavage boundaries. Founder cells that give rise exclusively to ectoderm or endoderm arise within the veg1 territory sometime after the eighth cleavage (the early blastula stage), but before the mesenchyme blastula stage, one to two cell divisions later (Ruffins and Ettensohn, 1996; Logan and McClay, 1997). The timing of founder cell formation corresponds closely to the stage at which prospective endoderm cells acquire the ability to differentiate autonomously *in vitro* (Chen and Wessel, 1996), although such *in vitro* experiments do not distinguish between veg2- and veg1-derived endoderm. As discussed later, it has been proposed that the definitive ectoderm–endoderm boundary is established via local cell interactions that take place at the late blastula stage and that the boundary is stably fixed before the onset of invagination (Logan and McClay, 1998).

A high-resolution fate map of the vegetal plate of the *L. variegatus* mesenchyme blastula has been generated by labeling single cells *in situ* with DiI (Ruffins and Ettensohn, 1996; Fig. 2). This territory of the blastula, identifiable as a thickened region of the epithelial wall at the vegetal pole, is of special importance for several reasons. Although it consists of only ~200 cells, or roughly one-quarter of the cells of the late blastula, the vegetal plate contributes more than half of the cell types of the larva, including all mesodermal and endodermal cell types. In addition, the cells of the vegetal plate execute the morphogenetic movements of gastrulation. Much of the vegetal plate consists of descendants of the veg2 layer (Hörstadius, 1973; Logan and McClay, 1997; Ransick and Davidson, 1998). Because there is typically a rather gradual transition between the thickened, vegetal epithelium and the thinner, lateral wall of the embryo, however, the correspondence between the lineage compartment and the morphological structure is only approximate.

The fate map shows that ~74 mesodermal precursors occupy the central region of the vegetal plate, surrounded by ~155 endodermal precursors. The latter are arranged in expanding concentric rings of prospective foregut, midgut, and hindgut. The initial inpocketing of the archenteron involves only prospective mesoderm cells, whereas the future endoderm (a ring 4–5 cells wide) involutes during gastrulation. This is consistent with numerous observations demonstrating limited involution during and after gastrulation (reviewed in Ettensohn, 1999). The fate map also reveals an oral–aboral (O-Ab) polarity in mesodermal cell fates—prospective pigment cells and blastocoelar cells are preferentially localized on the aboral and oral sides of the vegetal plate, respectively. O-Ab polarity of the vegetal plate has been demonstrated at the level of gene expression as well. In *S. purpuratus*, expression of *CyIIa* actin mRNA at the mesenchyme blastula stage is restricted to 10–15 presumptive secondary mesenchyme cells (SMCs) on the oral side of the plate (Miller *et al.*, 1996). In *L. variegatus*, apical localization of a Notch homolog is asymmetric along the O-Ab axis of the vegetal plate (Sherwood and McClay, 1997). These observations reveal that the O-Ab polarity of the embryo, shown previously to be expressed in the ectoderm and the ectoderm-dependent arrangement of the PMCs, is also evident in the vegetal plate and archenteron.

IV. Gene Expression Maps of the Early Sea Urchin Embryo

Antibody and cDNA probes have been essential tools in defining the progressive regionalization of the embryo. At present, the spatial patterns of expression of more than 120 different mRNAs have been examined by *in situ* hybridization. This number does not include homologs of genes that have been analyzed in more than one species. While this sample size is large enough to generate complex gene expression maps, it represents only about 1% of the mRNA species expressed during development (Davidson, 1989) and highly abundant transcripts are certainly overrepresented in the population. In addition to cDNA markers, the expression patterns of several hundred monoclonal antibodies have been examined through the efforts of many laboratories.

Most maternal mRNAs that have been examined are distributed uniformly in the unfertilized egg and early zygote. Several exceptions are known, however. The *bep1*, *bep3*, and *bep4* mRNAs, which encode a family a cell surface proteins related to the hyaline layer protein HLC-32, are restricted to the animal one-half to two-thirds of the unfertilized egg of *Paracentrotus lividus* (Di Carlo *et al.*, 1994; Montana *et al.*, 1996). An mRNA encoding a homolog of the vertebrate COUP-TF protein, an orphan steroid/thyroid hormone receptor family member, is expressed in a polarized fashion in the unfertilized eggs of at least three species along an axis perpendicular to the A-V axis (Vlahou *et al.*, 1996). Comparison between the positions of the transcripts, the first cleavage plane, and the O-Ab axis indicates that this early asymmetry does not align precisely with this axis, but is instead offset by 45° in two of the species examined. In *P. lividus*, a maternal mRNA encoding a fibrillar collagen (*COLL1α*) is distributed asymmetrically in the egg, also along an axis perpendicular to the A-V axis (Gambino *et al.*, 1997). The relationship between this polarized expression and the future oral–aboral or right–left axes of the larva has not been established.

Zygotic transcription in the sea urchin is detectable at the earliest stages of development, with the rate of RNA synthesis per cell increasing to a maximum by late cleavage (Davidson, 1986). The earliest, zygotically activated genes known to have spatially restricted patterns of expression are the VEB genes (see later) and the homeobox-containing gene *PlHbox 12* (DiBernardo *et al.*, 1995). Transcripts encoded by these genes begin to accumulate during early cleavage, are expressed at maximal levels at the morula stage ($\sim$128-cell stage), and disappear before gastrulation begins. *PlHbox12* mRNA is asymmetrically expressed as early as the 16-cell stage in certain mesomeres and macromeres along one side of the embryo, orthogonal to the A-V axis. The relationship between this axis and the major axes of polarity of the larva has not been determined. As discussed later, the VEB genes are expressed in a radially symmetrical pattern restricted to the animal two-thirds to three-fourths of the embryo. This "nonvegetal" pattern of expression has been reported to arise as early as the 16-cell stage (see Wei *et al.*, 1999a).

By the end of the blastula stage, the embryo has been partitioned into several domains of gene expression. Davidson and colleagues (Davidson, 1989; Davidson *et al.*, 1998) have identified five major territories of gene expression in the blastula: small micromere descendants, large micromere descendants, oral ectoderm, aboral ectoderm, and vegetal plate. This subdivision has been very useful in focusing attention on the molecular mechanisms that underlie the regionalization of the early embryo and on the possible relationships between early lineage compartments and domains of gene expression. As discussed later, however, the five-territory model may oversimplify the progressive regionalization of the early embryo. In addition, although it has been suggested that these domains of gene expression can be mapped directly onto early lineage compartments (Davidson, 1989; but see Davidson *et al.*, 1998, for a modified view), in only two cases, the small and large micromere territories, does this appear to be the case.

The large micromere descendants, the primary mesenchyme cells (PMCs), are unique in their expression of a battery of genes involved in skeletal morphogenesis (reviewed in Ettensohn *et al.*, 1997). The expression of the *SM50* gene, for example, has been analyzed in detail (Killian and Wilt, 1989). Using sensitive methods, *SM50* transcripts can be detected at low levels (~70 transcripts/embryo) during late cleavage. The spatial distribution of the mRNA at this early stage, however, is unknown. A major increase in transcription of the *SM50* gene occurs at the late blastula stage, concomitant with a sharp increase in mRNA level. As assayed by Northern blotting, the major phase of accumulation of *msp130* mRNA precedes that of *SM50*, and *in situ* hybridization shows that the *msp130* transcript is restricted to presumptive PMCs by the late blastula stage, prior to ingression (Harkey *et al.*, 1988; Guss and Ettensohn, 1997). Although the expression of *SM50*, *msp130*, and other mRNAs is modulated within the PMC population at later developmental stages (Harkey *et al.*, 1992; Guss and Ettensohn, 1997), no new cell types arise from this population. The large micromere–PMC lineage, therefore, clearly constitutes a distinct territory of gene expression by the late blastula stage.

The four small micromeres divide only once more during embryogenesis, producing eight progeny that contribute to the coelomic pouches (Pehrson and Cohen, 1986). The small micromere descendants constitute a territory defined primarily by the absence of gene products expressed in the remainder of the vegetal region at the blastula stage (Ransick *et al.*, 1993; Miller *et al.*, 1996, Sweet *et al.*, 1999). These cells have been shown to express high levels of β-catenin (Miller and McClay, 1997). At present, there is no evidence that the small micromeres diversify during later embryogenesis.

The aboral ectoderm is identified by several molecular markers, including *CyIIIa* and *CyIIIb* (Cox *et al.*, 1986), *Spec1/2* (Hardin et al., 1988; but see Angerer *et al.*, 1999), and arylsulfatase (Yang *et al.*, 1989). Sensitive methods have been used to detect *CyIIIa* transcription as early as the 60- to 120-cell stage, al-

1. Patterning the Early Sea Urchin Embryo

though the spatial distribution of the mRNA at early stages is not known (see Davidson *et al.*, 1998). When the spatial patterns of *CyIIIa* and *Spec1* can be detected by *in situ* hybridization at the late blastula stage, the mRNAs are enriched in the aboral ectoderm. Evidence shows that the aboral ectoderm is subdivided into smaller domains of gene expression later in development; in particular, cells at the posterior apex of the embryo are distinctive in their levels of expression of *EGFII* and *SpHbox1* mRNAs (Angerer *et al.*, 1989; Grimwade *et al.*, 1991). The boundary between the aboral ectoderm and the oral ectoderm, defined by the position of the ciliary band, is established late in gastrulation, apparently as a result of cell interactions at this interface (Cameron *et al.*, 1993; Davidson *et al.*, 1998).

There are very few markers for the oral ectoderm, and none are truly restricted to this region. The monoclonal antibody (mAb), EctoV (Coffman and McClay, 1990), has been the most widely used marker for this territory, although the molecule recognized by this antibody is also expressed in the foregut. The classification of the oral ectoderm as a discrete territory is further confounded by the fact that this region gives rise to several cell types, including the oral epithelium, neurons, and cells of the stomodeum. A striking, restricted expression of an *orthopedia* homolog has been described in a small subset of oral ectoderm cells of *P. lividus*, at gastrula and postgastrula stages (DiBernardo *et al.*, 1999). Angerer *et al.* (1990) have noted that while no mRNAs that are restricted to the oral ectoderm have been identified, several are expressed in both the oral ectoderm and the endoderm. These include transcripts encoding gene products associated with cell cycling/growth, such as cyclin B and histones, which become progressively restricted to these regions at later embryonic stages (Kingsley *et al.*, 1993). Based on all these considerations, in the absence of additional molecular markers, there is little justification for considering the oral ectoderm to be a discrete territory at any developmental stage.

The vegetal plate constitutes, at most, a very transient gene expression territory. As soon as the vegetal plate can be distinguished morphologically and the spatial expression of vegetal plate-specific gene products becomes detectable by *in situ* hybridization, this region of the embryo is partitioned into subterritories of gene expression. *CyIIa* transcripts are restricted to presumptive secondary mesenchyme cells (SMCs) on the oral side of the vegetal plate by the mesenchyme blastula stage (Miller *et al.*, 1996). By this same stage, the determinant recognized by mAb Sp1 is expressed specifically by prospective pigment cells (Gibson and Burke, 1985). Restricted expression of a SMC-specific determinant recognized by mAb SMC 1 can be detected at the hatched blastula stage, prior to PMC ingression (Sweet *et al.*, 1999). Notch protein expression is downregulated specifically in the prospective mesodermal region of the vegetal plate at an even earlier stage (Sherwood and McClay, 1997). Concomitant with the restricted expression of these markers, other molecules are more widely expressed throughout most or all of the thickened vegetal epithelium at the late blastula stage (Ransick *et al.*, 1993; Harada *et al.*, 1995, 1996; Miller *et al.*, 1996; Wang *et al.*, 1996). Evidence shows that the expression of one such marker, *Endo16*, may be confined transiently to cells of the veg2 lineage (Ransick and Davidson, 1998). On the whole, however, there seems little basis for the view that the vegetal plate represents a distinct molecular or lineage territory.

Maps of the known territories of gene expression in the early embryo are shown in Fig. 3. These maps are based on *in situ* hybridization and immunolocalization studies and are therefore limited by the sensitivity of those methods. It is evident that the specification events leading to these patterns of gene expression precede the patterns themselves. Indeed, there can be a considerable interval between the two. For example, as discussed later, cell isolation and transplantation studies indicate that the founder cells of the skeletogenic mesenchyme, the large micromeres, are determined when they arise at the 32-cell stage. One can infer that events sufficient to entrain the specific molecular program of this cell lineage have occurred by that time. The earliest stage at which *in situ* hybridization and immunolabeling studies provide unambiguous evidence of cell-type-specific expression of genes by the large micromere descendants, however, is the late blastula stage. One critical issue here is that the more sensitive methods used to detect the appearance of cell-type-specific transcripts at earlier (cleavage) stages have been applied only on a whole-embryo basis (e.g., Killian and Wilt, 1989). It would be useful to apply sensitive (e.g., polymerase chain reaction (PCR)-based) methods to isolated blastomeres of specific types in order to determine more precisely when *SM50*, *msp130*, and other transcripts first accumulate differentially in cells of specific lineages.

V. Mechanisms of Patterning Along the Animal–Vegetal Axis

A. Maternal Patterning

1. Evidence of Maternal Polarity

A variety of morphological, molecular, and developmental features reveal that the unfertilized egg is polarized along a primordial axis, the A-V axis. These include the position of the jelly canal and female pronucleus and the distribution of pigment granules and maternal mRNAs (Hörstadius, 1973; Schroeder, 1980a,b; DiCarlo *et al.*, 1994, 1996). When animal and vegetal fragments are isolated from unfertilized eggs and subsequently fertilized, they show stereotypical differences in developmental potential that correspond to the fate map (Hörstadius, 1973; Maruyama *et al.*, 1985). In other echinoderms, cytoplasmic depletion and transfer experiments show that vegetally localized determinants are responsible for the specification of vegetal cell fates, although similar studies have not yet been carried out with echinoid embryos (Kiyomoto and Shirai, 1993; Kuraishi and Osanai, 1994). These studies provide evidence that maternal determinants, still unspecified in nature, are distributed asymmetrically along the A–V axis.

2. Micromere Specification

The micromeres are the only blastomeres of the 16-cell embryo that show a highly restricted developmental potential. These cells divide unequally at the fifth cleavage division. The large daughter cells (large micromeres) are allocated to a skeletogenic fate whereas the small daughter cells (small micromeres) contribute to the coelomic pouches. The developmental potential of the micromeres has been analyzed in many types of cell isolation and transplantation experiments (Hörstadius, 1973; Okazaki, 1975; Livingston and Wilt, 1990; Khaner and Wilt, 1991; Ransick and Davidson, 1993). No conditions that prevent the cells from giving rise to skeletogenic mesenchyme have been found. Thus, the large micromeres are determined in the classical sense. Interestingly, the small micromeres show no skeletogenic potential when tested under a variety of conditions (Khaner and Wilt, 1991; Ettensohn and Ruffins, 1993).

Specification of the large micromere–PMC lineage is entrained by asymmetrically distributed maternal components and is associated with cortical changes in the early embryo. When unfertilized eggs are bisected into animal and vegetal halves and fertilized, only the vegetal half gives rise to skeletogenic mesenchyme (Hörstadius, 1973; Maruyama et al., 1985). Shortly before the cleavage, local actin-mediated changes in the cell cortex of the four vegetal blastomeres result in a displacement of the cortex toward the animal side of these cells (Schroeder, 1980; Tanaka, 1981). This is associated with a movement of the spindles toward the vegetal pole, probably through an interaction with a cortical attachment site (Dan and Tanaka, 1990; Holy and Schatten, 1991). The resulting unequal division produces the four micromeres. If the cortical displacement is inhibited by treating embryos with pharmacological agents, the fourth cleavage is equalized and no micromeres form (Tanaka, 1976; Langelan and Whiteley, 1985). Such embryos also fail to form early ingressing, skeletogenic mesenchyme.

The precise role of the unequal fourth cleavage in micromere determination is unknown. It may serve to sequester a vegetally localized (or vegetally activated) regulatory molecule above a critical threshold concentration. Even if a putative regulator was uniformly distributed along the A-V axis, the unequal division might increase the effective local concentration of the factor if it was specifically associated with the cortex or plasma membrane (Duncan et al., 1995). The function of a putative regulatory factor might be modulated by cell size in other ways. For example, the small volume of the micromeres would be expected to result in a lower molar ratio of a soluble transcription factor to DNA; this might create conditions under which the quantity of the DNA-binding factor becomes limiting. It has been proposed that such a mechanism might account for the distinctive bulk chromatin structure of the micromeres (Angerer and Angerer, 1997). As another possible consequence of a high nucleus-to-cytoplasm ratio, the micromeres would be expected to have a relatively large ratio of zygotically synthesized mRNA to maternal gene products (Senger and Gross, 1978). This might be significant if a balance between specific maternal and zygotic gene products affects skeletogenic determination.

Thus far it has not been possible to dissociate the cortical change from the unequal division experimentally. Therefore, it remains possible that the cortical change, and not the unequal division per se, plays a primary role in micromere determination. The local cortical change might alter the mobility or signaling properties of ligands on the micromere surface. The notion that micromeres have distinctive surface properties is supported by the finding that concanavalin A induces capping of its receptors on the surfaces of micromeres, but not on mesomeres or macromeres (Roberson et al., 1975).

Recent studies have implicated a member of the ETS class of transcriptional regulators in micromere specification. This family consists of more than 30 related proteins that share a highly conserved DNA-binding (ETS) domain (Sharrocks *et al.*, 1997; Wasylyk *et al.*, 1998).The ETS domains of the sea urchin *LvEts1* and human *ets-1* genes, for example, are identical at 81/84 amino acid positions. ETS proteins have been grouped into several subfamilies based on the sequence of the DNA-binding domain, the position of this domain within the protein, and additional sequences outside the DNA-binding domain that are shared by members of subfamilies. In combination with other proteins, ETS family members act as sequence-specific transcriptional activators or repressors.

An *ets-1/ets-2* homolog (*Hp-ets*) has been cloned from *Hemicentrotus pulcherrimus* and implicated in micromere/PMC specification (Kurokawa *et al.*, 1999). Whole-mount *in situ* hybridization shows that *Hp-ets* transcripts are expressed throughout the unfertilized egg and early embryo, but become restricted to presumptive PMCs at the blastula stage. Overexpression of *Hp-ets* mRNA causes the conversion of many cells of the embryo to a mesenchymal phenotype and results in upregulation of the skeletogenic marker gene, *SM50*. The 5' regulatory region of this gene contains a consensus Hp-ets binding site; mutations within this site substantially decrease the expression of a luciferase reporter gene. Overexpression of a putative dominant-negative form of Hp-ets lacking the N-terminal activation domain suppresses the formation of primary mesenchyme, expression of *SM50*, and skeletogenesis. Together, these findings constitute strong evidence that *Hp-ets* plays a role in skeletogenic differentiation.

This work raises a number of important questions, including the extent to which the full molecular program of skeletogenesis can be elicited by the overexpression of *Hp-ets,* the stage in the micromere specification pathway at which the protein functions, and whether maternal and/or zygotically produced *Hp-ets* has a key functional role. As discussed later, recent findings show that levels of nuclear β-catenin in the micromeres influence their specification. Although the underlying mechanism is unknown, one speculation is that β-catenin might regulate micromere fate by modulating the activity, localization, or expression of the Hp-ets protein.

3. β-Catenin and Early Patterning

In many metazoans, β-catenin plays a central role in specifying the fates of early blastomeres (see Miller and Moon, 1996; Moon *et al.*, 1997; Han 1997; Zeng *et al.*, 1997). Among deuterostomes, the possible role of this pathway in early blastomere interactions and axis specification has been studied most extensively in *Xenopus* (Harland and Gerhart, 1997; Moon *et al.*, 1997). Maternal β-catenin is both necessary and sufficient for the specification of dorsal cell fates and the establishment of the secondary (D-V) axis. Overexpression of a variety of positive regulators of the Wnt/β-catenin pathway on the ventral side of the embryo causes the formation of an ectopic dorsal axis, whereas overexpression of negative regulators on the dorsal side generally ventralizes embryos. β-catenin accumulates in the nuclei of cells on the dorsal side of the embryo as early as the 8-cell stage (Larabell *et al.*, 1997). The prevailing view is that the role of this pathway in early patterning is in establishing the dorsal–vegetal (Nieuwkoop) signaling center, which in turn induces overlying cells to form Spemann's organizer (Harland and Gerhart, 1997). β-Catenin has been shown to be one of several factors that interact with members of the TCF/LEF-1 family of DNA-binding proteins to regulate the transcription of downstream genes (Molenaar *et al.*, 1996; Roose *et al.*, 1998; Waltzer and Bienz, 1998). Considerable data support the view that this is the mechanism by which the Wnt/β-catenin pathway regulates cell fates, although Wnt proteins and β-catenin have also been shown to affect gap junction-mediated communication in a transcription-independent fashion (Olson *et al.*, 1991; Krufka *et al.*, 1998).

There is evidence that in *Xenopus*, the nuclear localization of β-catenin is regulated in a ligand-independent manner, circumventing secreted Wnt proteins and their receptors. Wnts and *dishevelled* both increase signaling through the pathway and cause secondary axis formation if expressed on the ventral side of the embryo; this ectopic axis can be blocked by coexpressing reagents that are expected to interfere with Wnt and *dishevelled* activity. Surprisingly, these same reagents do not block formation of the endogenous axis when expressed dorsally (Hoppler *et al.*, 1996; Sokol, 1996; Leyns *et al.*, 1997; Wang *et al.*, 1997). It has therefore been speculated that the cortical rotation at fertilization may activate the pathway downstream of Wnts/frizzled/dishevelled in a cell-autonomous manner. One model is that a key control point is GSK3, a serine/threonine kinase that phosphorylates β-catenin. Phosphorylation by GSK3 targets β-catenin for rapid proteolysis via the ubiquitin/proteosome pathway (Aberle *et al.*, 1997; Orford *et al.*, 1997; Laney and Hochstrasser, 1999). It has been proposed that the cortical rotation at fertilization might redistribute an inhibitor of GSK3 activity to the dorsal side of the zygote (Yost *et al.*, 1998). Although this is one plausible model, the phosphorylation/degradation of β-catenin is potentially subject to complex regulatory control. GSK3 activity is regulated both positively and negatively by phosphorylation at multiple sites (see van Weeren *et al.*,

1998). GSK3 and β-catenin can exist in a complex with several other proteins, including axin, adenomatous polyposis coli protein (APC), GSK3-binding protein (GBP), the B56 subunit of protein phosphatase 2A, and probably other proteins (Hart *et al.*, 1998; Ikeda *et al.*, 1998; Yost *et al.*, 1998; Seeling *et al.*, 1999). These various proteins can regulate the phosphorylation of β-catenin by GSK3 in a positive (e.g., axin) or negative (e.g., GBP) fashion. In addition to these known regulators of GSK3 activity, it is also possible that the availability or activity of components of the ubiquitin/proteosome pathway might regulate the rate of β-catenin degradation (Laney and Hochstrasser, 1999). At present, whether any of these potential regulators are differentially expressed or activated along the dorsoventral axis of the embryo is unknown.

In the sea urchin, studies from the McClay, Gache, and Klein laboratories have provided compelling evidence that components of the Wnt/β-catenin signaling pathway are involved in the specification of cell fates along the A-V axis. During cleavage, endogenous β-catenin becomes concentrated in the nuclei of vegetal cells in a dynamic, stereotypical pattern (Logan *et al.*, 1999). At the 16-cell stage, the protein is concentrated in the nuclei of micromeres and macromeres. When the veg1 and veg2 layers form at the 60-cell stage, nuclear β-catenin is restricted to the veg2 cells. Later in development, nuclear localization diminishes in the veg2 cells and micromeres, whereas it increases in prospective endoderm cells derived from the veg1 layer.

Signaling through the pathway is both necessary and sufficient for the expression of mesodermal and endodermal fates. Thus, embryos are vegetalized by overexpression of positive regulators of the pathway, including sea urchin or *Xenopus* mRNAs encoding (1) a hyperstable β-catenin with mutated GSK3 phosphorylation sites (Wikramanayake *et al.*, 1998), (2) a dominant-negative form of GSK3 inactivated by a single amino acid substitution in the ATP-binding site (Emily-Fenouil *et al.*, 1998), or (3) a TCF/LEF-1 variant in which the β-catenin-binding domain has been replaced with the VP16 activation domain (Vonica *et al.,* 2000). Such embryos closely resemble those that have been treated with the classical vegetalizing agent LiCl, which as been shown to act as an inhibitor of GSK3 (Klein and Melton, 1996; Stambolic *et al.*, 1996). Conversely, the formation of mesoderm and endoderm is suppressed by overexpression of negative regulators of the pathway, including (1) wild-type GSK3 (Emily-Fenouil *et al.*, 1998), (2) *Xenopus* C-cadherin (Wikramanayake *et al.,* 1998b) or the cytoplastic domain of sea urchin LvG-cadherin (Logan *et al.*, 1999), or (3) or TCF/LEF-1 protein, which lacks the β-catenin-binding domain (Vonica *et al.,* 2000). It is of interest that overexpression of a dominant-negative XWnt-8 does not block early vegetal plate specification, but affects later patterning events during gastrulation (Wikramanayake *et al.*, 1998a). This suggests that, as in *Xenopus*, the pathway may be regulated in a Wnt-independent fashion during early development, whereas Wnt-mediated signaling becomes important during gastrulation (Sokol, 1996).

What is the normal role of the β-catenin pathway in patterning? Because the micromeres have an important function in inducing mesoderm and endoderm

1. Patterning the Early Sea Urchin Embryo

(see later), alterations in β-catenin levels might be acting by perturbing micromere signaling. Indeed, it seems clear that the animalized phenotype observed following downregulation of β-catenin levels arises, at least in part, by this mechanism. Embryos that have been fully animalized by overexpression of GSK3 or cadherin form no skeletogenic mesenchyme, even though they produce micromeres at the 16-cell stage (Emily-Fenouil *et al.*, 1998; Wikramanayake *et al.*, 1998b; Logan *et al.*, 1999), indicating that micromere specification is perturbed in such embryos. Direct evidence that the signaling properties of micromeres are also affected comes from the finding that micromeres isolated from embryos injected with the cytoplasmic domain of sea urchin LvG-cadherin have a reduced ability to induce secondary axes when recombined with animal blastomeres (Logan *et al.*, 1999). It is not known whether the converse is true, i.e., whether micromeres isolated from embryos vegetalized by overexpression of β-catenin or dominant-negative GSK3 have more potent inductive properties than normal micromeres. These data suggest that the expression of signaling factors by micromeres is regulated in a transcription-dependent manner, through the action of β-catenin on TCF/LEF-1 and downstream target genes. If this view is correct, the emergence of signaling properties on the part of the micromeres would require sufficient time for the transcription of such target genes, the translation of the encoded proteins, and their delivery to the micromere surface.

Despite these findings, it seems unlikely that β-catenin functions only to regulate micromere signaling. Several lines of evidence argue, instead, that β-catenin levels can directly regulate the fates of early blastomeres in a cell-autonomous fashion: (1) After the removal of micromeres, embryos form both endoderm (Ransick and Davidson, 1995) and mesoderm (Sweet *et al.,* 1999), although to a variable extent and in a delayed fashion. Embryos animalized by overexpression of cadherins or GSK3, however, show a much more severe phenotype and completely lack mesoderm and endoderm. This suggests that downregulation of β-catenin levels in macromeres affects their specification in a cell-autonomous fashion. (2) Wikramanayake *et al.* (1998b) have shown that animal blastomeres isolated from 8-cell stage embryos overexpressing hyperstable β-catenin are vegetalized. Isolated mesomeres can also be vegetalized by treatment with LiCl (Livingston and Wilt, 1989), an inhibitor of GSK3. These data indicate that levels of β-catenin regulate the specification of animal cells by a mechanism independent of micromere signaling. (3) Overexpression of wild-type or kinase-deficient GSK3 causes an expansion or contraction, respectively, of the domain of expression of the hatching enzyme (*HE*) gene (Emily-Fenouil *et al.*, 1998). Coupled with the evidence that *HE* expression is regulated in a cell-autonomous fashion (see later), this finding suggests that GSK3 acts directly within cells to regulate their fate. Collectively, these findings suggest that components of the GSK3/β-catenin pathway are part of a maternally derived system that acts autonomously in early blastomeres.

Davidson *et al.* (1998) have proposed a "parallel input" model that suggests that the *cis*-regulatory elements of genes expressed in the endomesoderm integrate information from both β-catenin/HMG-box (TCF/LEF-1) inputs and micromere-dependent signals. They propose that early interactions between β-catenin/HMG complexes and the *cis*-regulatory elements of endomesodermal genes increase the subsequent responsiveness of such genes to transcription factors whose activities are micromere signal-dependent. This model offers a plausible explanation for the surprising finding that activation of endomesodermal genes can occur without appreciable levels of nuclear β-catenin. Micromeres transplanted to the animal pole of early cleavage stage embryos can induce both mesodermal and endodermal cell types, even though the animal cells do not express detectable levels of nuclear β-catenin (Logan *et al.*, 1999; see later). The parallel input model suggests that nuclear β-catenin, while not absolutely required for endomesoderm specification, sensitizes cells to micromere-derived signals. Some evidence exists to support this model; macromeres have been shown to be far more sensitive than mesomeres to the micromere-derived, mesoderm-inducing signal (Sweet *et al.*, 1999) and their competence to respond to the signal is suppressed if β-catenin levels are downregulated in the cells (D. R. McClay, personal communication).

The finding that transplantation of micromeres to the animal pole induces the formation of a secondary archenteron without causing a detectable increase in nuclear β-catenin argues that nuclearization is not triggered by micromere induction. This suggests that during normal development, the accumulation of β-catenin in the nuclei of macromeres and their progeny during early cleavage may be a consequence of cell-autonomous processes. Consistent with this view, nuclearization of β-catenin appears to occur normally in macromeres following micromere removal and under culture conditions that prevent normal blastomere contacts (Logan *et al.*, 1999).

Although the most conspicuous effects of perturbing the β-catenin pathway are on A-V patterning, this pathway affects secondary (O-Ab) polarity as well. Embryos animalized by overexpression of C-cadherin show a partially oralized phenotype (Wikramanayake *et al.*, 1998b). The fact that both A-V polarity and O-Ab polarity are perturbed in such embryos supports the view that the β-catenin pathway links patterning along these two axes. In support of this view, overexpression of *Xenopus* pt (hyperstable) β-catenin in animal caps isolated from 8-cell embryos, or treatment of animal caps with LiCl, reestablishes polarity of the ectoderm along the O-Ab axis (Livingston and Wilt, 1990; Wikramanayake *et al.*, 1995, 1997, 1998b). At high enough concentrations, both treatments also cause a morphological vegetalization of animal caps, including the formation of an archenteron. It seems likely, therefore, that the rescue of the O-Ab axis in animal caps by β-catenin or LiCl is a secondary consequence of entraining vegetal-to-animal signaling processes, which are known to be required normally to polarize the ectoderm along that axis. In apparent contrast, it has been shown that low concentrations of LiCl can polarize animal caps without inducing detectable levels of endoderm-specific genes, implying a direct polarizing effect on the presumptive ectoderm (Wikramanayake *et al.*, 1997). It remains possible, however, that elevation of β-catenin levels confers a more vegetal fate (perhaps a veg1-

like fate) on some animal cap cells, causing them to transmit signals that pattern the ectoderm (see Wikramanayake *et al.*, 1998b). To partly resolve this issue it would be informative to determine whether signaling between blastomere layers is entrained in animal caps by treatment with low concentrations of LiCl or following injection with low doses of pt β-catenin.

4. Early Differential Gene Expression along the A-V Axis: The VEB Genes

Four genes that are transiently expressed at the early blastula stage have been identified. These genes, known collectively as the VEB (very early blastula) gene set, are among the first to be activated zygotically, suggesting that their expression is regulated directly by maternal factors. Two of these genes encode metalloendoproteases—*HE* and a BMP-1/tolloid homolog—whereas the others represent novel sequences. VEB mRNAs accumulate in a radially symmetrical domain in the animal region of the embryo but are not expressed at the vegetal pole (Lepage *et al.*, 1992a,b; Reynolds *et al.*, 1992; Angerer and Angerer, 1997). The domain of expression appears to vary among species and has been variously described as the animal-most two-thirds of the blastula, including only the prospective ectoderm (Lepage *et al.*, 1992b), or a somewhat larger region of the embryo that includes variable amounts of prospective endoderm and secondary mesenchyme (Reynolds *et al.*, 1992). Within this domain of expression, *HE* transcripts are expressed in a gradient with highest levels of expression at the animal pole (Reynolds *et al.*, 1992).

Several observations indicate that the initial activation of the VEB genes is regulated in a cell autonomous fashion. Activation of the VEB genes occurs on schedule even when embryos are cultured under conditions that interfere with normal blastomere contacts (Ghiglione *et al.*, 1993). As discussed later, a major signaling event in the early embryo is the induction of mesoderm and endoderm by the micromeres. The domain of *HE* expression, however, is completely insensitive to micromere-derived signals (Ghiglione *et al.*, 1996). The boundary of expression can be shifted toward the animal pole, however, by LiCl, an inhibitor of GSK3, or by overexpression of a kinase-inactive GSK3 (Emily-Fenouil *et al.*, 1998). Conversely, overexpression of wild-type GSK3 shifts the boundary toward the vegetal pole. The early, autonomous activation of the VEB genes along the prefertilization (A-V) axis suggests that the expression of these genes is controlled by a maternal system that regulates levels of β-catenin along the axis.

The *cis*-regulatory structure of the *HE* gene has been examined in two species, *S. purpuratus* (Wei *et al.*, 1995, 1997a,b) and *P. lividus* (Ghiglione *et al.*, 1997). These studies have revealed multiple (9–12) sites that interact with several different regulators. No negative spatial regulatory sites have been identified in the *S. purpuratus* gene, whereas there is evidence for negative spatial regulation in *P. lividus*. Some sites are present in both species (e.g., multiple CCAAT sites) whereas others are not. No single *cis* element is critical, although inactivation of multiple sites leads to reduced expression. Surprisingly, the *cis*-regulatory region of another VEB gene, the BMP-1-like protease, shows little similarity to the *HE* gene and appears to interact with a different set of DNA-binding proteins (Kozlowski *et al.*, 1996). It has been shown that the sharp decline in *HE* mRNA levels at the late blastula stage is not due to the disappearance of positively acting factors that regulate the gene, but may instead be associated with changes in chromatin structure (Wei *et al.*, 1997b). The downregulation of *HE* mRNA levels is also regulated posttranscriptionally through a decrease in mRNA stability (Ghiglione *et al.*, 1993). Unlike the transcriptional activation and repression of the gene, downregulation is dependent on cell signaling events (Ghiglione *et al.*, 1993).

The factors that regulate VEB expression are likely to be closely linked to the maternal axis and the β-catenin patterning system. One positive regulator of the *HE* gene in *S. purpuratus* has been identified by yeast one-hybrid screening and found to be a member of the ETS family (Wei *et al.*, 1999b). This protein, SpEts4, is encoded by maternal transcripts that are distributed uniformly in the egg and cleavage stage embryo. During the blastula stage, the maternal transcripts are replaced by zygotic mRNA that accumulates in a nonvegetal pattern congruent with *HE* gene expression (Wei *et al.*, 1999a). Thus, the zygotic expression of *SpEts4* appears to be downstream of the maternal polarity system and may be under the same ensemble of controls as its target gene. A direct link between maternal A-V polarity and VEB gene expression has therefore not yet been established.

B. Cell–Cell Interactions

1. Micromere Signaling

Signals emanating from the vegetal region of the embryo during cleavage are important in specifying blastomere fates. Signals from the micromeres are both necessary and sufficient for the normal specification of overlying cells (macromeres and their descendants) as mesoderm and endoderm. Microsurgical removal of the micromeres leads to a reduction in the number of cells that express *Endo16*, an endoderm marker, and causes a delay in gastrulation (Ransick and Davidson, 1995). Recombination of micromeres with animal pole blastomeres induces the ectopic expression of endodermal markers and the formation of a secondary gut (Hörstadius, 1973; Khaner and Wilt, 1991; Ransick and Davidson, 1993). Other gene products expressed in the prospective endoderm, such as *SpKrox1* (Wang *et al.*, 1996), *forkhead* homologs (Harada *et al.*, 1996; Luke *et al.*, 1997), *LvN1.2* (Wessel *et al.*, 1989), and *Endo1* (Wessel and McClay, 1985), are also likely to be downstream of this signaling pathway.

Similarly, micromeres can induce animal blastomeres to give rise to mesoderm. Micromere signaling induces mesomeres to give rise to pigment cells (Khaner and Wilt, 1991) and cells that express a skeletogenic fate upon the removal of PMCs, a property normally exhibited by SMCs (Ettensohn and McClay, 1988; Minokawa *et al.*, 1997). New monoclonal antibody markers for nonskeletogenic mesoderm, SMC 1 and SMC 2, have allowed a more detailed analysis of the role of micromere signaling in mesoderm formation (Sweet *et al.*, 1999). Removal of micromeres at the 16-cell stage blocks the normal formation of nonskeletogenic mesoderm in the vegetal plate of the blastula. Micromeres can also induce the formation of SMC 1- and SMC 2-positive mesoderm cells from animal caps isolated from 8-cell stage embryos, although in numbers considerably lower than normally found in the vegetal plate. Thus, micromere-derived signals are necessary for the normal specification of nonskeletogenic mesoderm and are sufficient to induce these cells types ectopically in mesomere derivatives.

Although micromere induction is required for the normal formation of both mesoderm and endoderm, there is also evidence that these tissues are specified in part by cell-autonomous mechanisms. Even when micromeres are removed after the fourth cleavage division, variable numbers of cells in the vegetal plate still express *Endo16* on schedule and the embryos invaginate after a delay (Ransick and Davidson, 1995). Some (but not all) mesodermal cell types also arise, albeit in reduced numbers and after a delay of several hours (Sweet *et al.*, 1999). One explanation for these findings is that even very brief micromere signaling is sufficient to induce endoderm and mesoderm. In *Caenorhabditis elegans*, for example, critical blastomere interactions take place on a time scale of minutes (Schnabel and Priess, 1997). In the sea urchin, technical constraints prevent surgical removal of the micromeres earlier than 10 min after their formation, which may be sufficient time for induction to occur (Ransick and Davidson, 1995). An alternative explanation, however, is that the macromeres have a limited capacity to give rise to mesoderm and endoderm in the absence of micromere-derived cues. Support for this view comes from animal blastomere isolation experiments. Individual animal blastomeres isolated at the 8-cell stage, prior to micromere formation, give rise to skeletogenic mesoderm and/or endoderm in almost 30% of the cases (Henry *et al.*, 1989). Moreover, the ability of animal blastomeres to form endoderm and mesoderm is correlated with the position of the third cleavage plane, which is naturally somewhat variable. Subequatorial cleavage leads to an increased frequency of mesoderm and endoderm formation (Henry *et al.*, 1989). Experimentally shifting the third cleavage plane toward the vegetal pole also increases the likelihood that animal cells isolated from 8-cell stage embryos will form mesoderm and endoderm (Kitajima and Okazaki, 1980). These findings have been interpreted to mean that maternal determinants involved in mesoderm/endoderm formation are not tightly localized at the vegetal pole and can be incorporated into animal blastomeres at significant levels, particularly if the third cleavage plane is subequatorial (Henry *et al.*, 1989). They do not formally exclude the possibility that vegetal-to-animal signaling occurs at the early 8-cell stage and that the strength of the vegetally derived signal is increased by a subequatorial cleavage. It is also possible that a signaling center is established by regulative mechanisms in isolated animal fragments and that formation of such a signaling center is enhanced by a subequatorial third cleavage. Nevertheless, these experiments show clearly that isolated mesomeres can give rise to endoderm and mesoderm in the absence of micromere-derived cues. It seems reasonable, then, to propose that macromeres have a similar capacity to form these tissues in a micromere-independent fashion.

The limited ability of macromeres to give rise to endoderm after micromere removal has been ascribed to the extensive regulative abilities of the embryo and has been considered distinct from the normal process of endoderm specification (Ransick and Davidson, 1995). A different interpretation, however, is that this behavior reveals the autonomous capacity of blastomeres to form mesoderm/endoderm independent of micromere signaling and is a consequence of normal specification processes that pattern the embryo along the A-V axis. Micromere formation and the appearance of an early ingressing, skeletogenic mesenchyme probably represent evolutionary modifications of an ancestral echinoderm developmental program (Schroeder, 1981; Wray and McClay, 1989). The autonomous mesoderm/endoderm-forming potential on the part of macromeres may be a persistent feature of a more ancient system of fate specification. Some light would be shed on this issue through a comparative analysis of fate specification mechanisms in cidaroid sea urchins and other echinoderms (e.g., Kiyomoto and Shirai, 1993; Kuraishi and Osanai, 1994).

Micromeres appear to transmit signals during a relatively broad window of developmental time. Removal of micromeres at progressively earlier stages between the fourth and sixth cleavage divisions results in a corresponding reduction in the numbers of vegetal plate cells that express *Endo16*, indicating that signaling during early cleavage is required for normal endoderm specification (Ransick and Davidson, 1995). Even when micromeres are allowed to remain in contact with macromeres until after the sixth cleavage division, however, the number of *Endo16*-expressing cells is usually lower than normal. This suggests that micromere-derived signals later in development may also be required for complete endoderm specification. Minokawa and Amemiya (1999) have found that micromeres and their early progeny have a relatively weak capacity to induce animal caps to form endoderm. If the micromeres have been allowed to mature *in vitro* to the equivalent of the blastula stage, however, even a brief (2 hr) interaction between animal caps and micromeres is sufficient to induce guts in a high proportion of cases. Minokawa and Amemiya (1999) note that their data and those of Ransick and Davidson (1995) can be reconciled if the signaling strength of the micromeres increases progressively during early development. The macromeres might be responsive to low levels of signal that are present during early development, whereas the mesomeres require higher levels only attained later in development. A more cumbersome model is that two qualitatively different signals are sent by early and late micromeres and that macromeres and mesomeres are sensitive to the early and late signals, respectively.

2. Downstream of Micromere Signaling: The Notch Pathway and Mesoderm Specification

The Notch receptor has been implicated in numerous cell fate decisions in vertebrates and invertebrates (reviewed by Kimble and Simpson, 1997; Artavanis-Tsakonas *et al.,* 1995). A homolog of the Notch receptor, LvNotch, has been cloned from *Lytechinus variegatus* (Sherwood and McClay, 1997). LvNotch is expressed uniformly along all cell surfaces during early cleavage, but is downregulated at the vegetal pole at the early blastula stage. At the mesenchyme blastula stage, expression increases on the apical surface of prospective endodermal cells, whereas the prospective nonskeletogenic mesoderm (NSM) in the center of the vegetal plate continues to lack LvNotch expression. Overexpression of activated LvNotch results in an increase in the number of NSM cells at the expense of prospective endoderm (Sherwood and McClay, 1999). Conversely, overexpression of a dominant-negative form of the receptor dramatically suppresses NSM formation, transfating these cells to endoderm. These findings provide strong evidence that Notch signaling has a central role in NSM fate specification and the positioning of the mesoderm/endoderm boundary. It has been suggested that the loss of apical LvNotch expression in the central region of the vegetal plate might reflect an internalization/degradation of LvNotch associated with active signaling through the pathway, consistent with the phenotypes of embryos overexpressing activated or dominant-negative receptors (Sherwood and McClay, 1999).

Sweet *et al.* (1999) and McClay (personal communication) have found that removal of the micromeres from 16-cell stage embryos blocks the downregulation of apical LvNotch expression in the central region of the vegetal plate. This significant finding links micromere signaling to the Notch pathway and provides the first evidence of a molecular pathway activated downstream of micromere signaling. A similar effect on the Notch expression pattern is seen in embryos overexpressing a dominant-negative LvNotch (Sherwood and McClay, 1999). In addition, micromere removal and disruption of Notch signaling both result in a delay and reduction in the formation of nonskeletogenic mesoderm (Sherwood and McClay, 1999; Sweet *et al.,* 1999). These findings indicate that β-catenin is required for the acquisition of signaling properties by micromeres, which in turn are required for the downregulation of apical Notch in the central region of the vegetal plate at the blastula stage and the specification of nonskeletogenic mesoderm in that region (Logan *et al.,* 1999; Sherwood and McClay, 1999; Sweet *et al.,* 1999). The situation may be different with respect to the endoderm, however. Micromere signaling is required for the normal specification of this germ layer (Ransick and Davidson, 1995), but disruption of Notch signaling does not appear to have a significant effect on endoderm formation (Sherwood and McClay, 1999). This suggests that endoderm induction by micromeres occurs via a Notch-independent mechanism.

Other key components of the Notch signaling pathway have also been identified in the sea urchin, including two proteins that process Notch (presenilin I and TNFα-converting enzyme), kuzbanian (a protease implicated in the processing of Delta), and Delta (Ettensohn, unpublished observations).

3. A Vegetal-to-Animal Signaling Cascade?

Davidson (1989) proposed that micromere signaling entrains a vegetal-to-animal cascade of signaling in the cleavage stage embryo. At present, the evidence in favor of this model is considerable, but indirect. Transplantation of micromeres to the animal pole not only induces neighboring cells to form a secondary gut, but also results in the formation of supernumerary PMC target sites within the ectoderm, several cell layers away from the donor cells (Hörstadius, 1973; Ransick and Davidson, 1993). This shows that micromere signaling can lead to global repatterning of the ectoderm, either by a signal acting over a distance or by a cell–cell relay mechanism. Macromeres and/or their derivatives (veg1 or veg2 cells) can also induce PMC target sites in the ectoderm (Benink *et al.,* 1997). Veg2 cells, like micromeres, can also induce mesomeres to form endoderm (Hörstadius, 1973; Logan and McClay, 1998). Thus, it is firmly established that macromere-derived cells have potent signaling properties that are similar in some respects to those of micromeres. As yet, however, the signaling properties of macromere descendants have not yet been shown to be dependent on, or enhanced by, prior interactions with micromeres, as would be required by the cascade model. Instead, the signaling properties of the macromere derivatives may arise as part of an autonomous program of specification (see earlier discussion).

If the micromeres entrain a vegetal-to-animal cascade, the next level of signaling would presumably be from veg2 to veg1. Veg1 cells normally give rise to ectoderm and variable amounts of endoderm (primarily midgut and hindgut). The extent to which veg1 cells *require* veg2 signaling to give rise to endoderm is unclear; however, it generally seems true that the veg1 layer shows a reduced ability to give rise to endoderm in the absence of more vegetal cells, and the endoderm that forms is poorly patterned (Hörstadius, 1973; Logan and McClay, 1998). As noted by Logan and McClay (1998), because cell isolation/recombination studies are carried out after the veg1 tier forms at the 60-cell stage, one cannot distinguish whether the ability of veg1 cells to form endoderm in the absence of more vegetal cells is a consequence of (1) an autonomous program of endoderm formation in the veg1 cells, (2) a long-range signaling from micromeres, or (3) a rapid transfer of veg2-derived cues. Because veg2 cells clearly have endoderm-inducing capacity (see earlier discussion) and because veg1 cells are at least partly dependent on vegetally derived signals to fully express their endoderm-forming potential, it seems probable that the veg2 cells transmit inductive signals that contribute to the patterning of the veg1 layer. The veg2–veg1 interaction must take place after the formation of the veg2 layer at the 60-cell stage and is apparently complete by the mesenchyme blastula stage, when the ectoderm–endoderm boundary within the veg1 territory seems to be stably fixed (Logan and McClay, 1998). The interaction may be concomitant with the segregation of veg1 progeny to the ectoderm or endoderm; in *L. variegatus*, this allocation occurs after the 9th or 10th cleavage divisions (i.e., at the late blastula stage) (Ruffins and Ettensohn, 1996; Logan and McClay, 1998).

Analysis of molecular expression patterns in the vegetal region of the embryo has revealed a progressive expansion of the initial domains of expression of three endodermal markers: alkaline phosphatase (AP), *Endo16*, and *SpKrox-1*. In starfish, expression of alkaline phosphatase (AP) occurs in two phases (Kurashi and Osanai, 1994). Early in gastrulation, AP activity is restricted to the anterior region of the archenteron (the future esophagus and anterior stomach). Only after archenteron elongation is complete does activity spread to the adjacent posterior region (the future posterior stomach and hindgut). AP is also expressed specifically in the gut in sea urchins, although the dynamics of its pattern of expression have not been analyzed. In *S. purpuratus*, the domain of *Endo16* mRNA expression expands into the veg1 territory during gastrulation (Ransick and Davidson, 1998). A similar expansion has been reported for *SpKrox-1*, but it precedes that of *Endo16* (Davidson *et al.*, 1998). The vegetal-to-animal expansion of the expression domains of these markers is likely to be a manifestation of cell interactions in the vegetal hemisphere, perhaps as a consequence of the veg2-to-veg1 signal. The underlying cell–cell signaling events must precede detectable changes in transcript levels or enzymatic activity, which are not demonstrable until after the beginning of gastrulation.

4. Interactions among Mesomere Derivatives

Intact caps of animal blastomeres isolated from 8- or 16-cell stage embryos generally differentiate as ectodermal spheroids (Hörstadius, 1973; Henry *et al.*, 1989). Single animal blastomeres isolated from 8-cell stage embryos, or pairs of mesomeres isolated from 16-cell stage embryos, each representing one-fourth of the animal hemisphere, exhibit a markedly increased capacity to give rise to skeletogenic mesoderm and endoderm (Henry *et al*, 1989). This finding suggests that homotypic interactions among mesomeres suppress an inherent mesoderm/endoderm-forming potential while reinforcing ectodermal specification. This restriction appears to be largely complete by the 32-cell stage, as quartets of animal blastomeres isolated at that stage have little ability to express nonectodermal fates (Henry *et al.*, 1989). The molecular basis of this interaction is currently unknown. One possibility, however, is that mesomere–mesomere interactions have the effect of activating or reinforcing the molecular pathways that prevent β-catenin from entering the nuclei of animal blastomeres during early cleavage.

VI. Mechanisms of Patterning Along the Oral–Aboral Axis

A. Expressions of the Oral–Aboral Axis

The secondary axis is established during oogenesis in several organisms, including some arthropods and brachiopods, but it can also be established during cleavage, as in mollusks and annelids (Goldstein and Freeman, 1997). In some organisms, rearrangements of the cytoskeleton following fertilization are involved in establishing this axis. This is especially common in the chordates, where the cue for cytoskeletal reorganization appears to be either the site of sperm entry or gravity (Eyal-Giladi, 1997). Molecular pathways involved in secondary axis formation are best understood in *Drosophila* and amphibian embryos (reviewed in Morisato and Anderson, 1995; Harland and Gerhart, 1997).

In the sea urchin, the first morphological asymmetry along the O-A axis appears at the late blastula stage, when bilateral regions of ectoderm become thickened on the oral side of the embryo (Okazaki *et al.*, 1962). It is at these sites that PMCs later form clusters and initiate skeletogenesis. The late blastula is also the earliest stage at which polarized expression of gene markers along the O-A axis has been demonstrated unambiguously by *in situ* hybridization (Angerer *et al.*, 1990; Miller *et al.*, 1996). Numerous morphological and molecular indicators of O-Ab polarity arise subsequently, as indicated in Figs. 1–3.

The nature and timing of the initial cue(s) that establishes the O-Ab axis are unknown. The finding that COLL1α and *SpCOUP-TF* transcripts are asymmetrically distributed within oocytes orthogonal to the A-V axis suggests that a secondary axis is established in some form during oogenesis (Vlahou *et al.*, 1996; Gambino *et al.*, 1997). During early cleavage, there arise additional molecular indices of polarization orthogonal to the A-V axis, although the relationship between such markers and the definitive O-Ab axis is ambiguous. As early as the 8-cell stage, a gradient of cytochrome oxidase activity can be detected with the region of maximum staining hypothesized to occur at the future oral pole (Czihak, 1971). As noted earlier, zygotic *PlHbox12* transcripts are expressed in a polarized manner perpendicular to the A-V axis during early cleavage (Di Bernardo *et al.*, 1995).

The relationship between the position of the first cleavage plane and the O-Ab axis varies among sea urchin species (reviewed by Jeffery, 1992; Henry, 1998). In some species, there is a fixed relationship between the cleavage plane and the axis (*S. purpuratus*, Cameron *et al.*, 1989; Henry *et al.*, 1992; *Heliocidaris tuberculata, Strongylocentrotus droebachiensis, Lytechinus pictus, Heliocidaris erythrogramma*, Wray and Raff, 1989; Henry *et al.*, 1992; and *Holopneustes purpurescens*, Morris, 1995). Depending on the species, the first cleavage plane can separate the embryo into right–left, oral–aboral, or oblique halves. In other species, the O-Ab axis can form at any angle relative to the first cleavage plane (*Paracentrotus lividus*, Hörstadius, 1973; *Hemicentrotus pulcherrimus*, Kominami, 1988; *Clypeaster rosaceus*, Henry *et al.*, 1992; and *L. variegatus*, Ruffins and Ettensohn, 1996; Summers *et al.*, 1996).

The tight correlation between the position of the first cleavage plane and the O-Ab axis in some species indicates that, at least in those cases, the cytoskeletal architecture of the 1 cell zygote is linked to the future axis. In the direct-developing species *H. purpurescens*, there is evidence that sperm entry polarizes a cytoskeletal reorganization, which predicts the future O-Ab axis (Morris, 1995). Studies on indirect developing sea urchins, however, are inconclusive in this regard (reviewed in Hörstadius, 1973). It is not clear whether the sperm provides any cue that aids in positioning the O-Ab axis, and there is no evidence yet of polarized cytoplasmic rearrangements following fertilization. In *L. variegatus*, the movement of the sperm aster within the egg appears to be involved in establishing the position of the first cleavage plane (Schatten, 1981), but not the position of the O-Ab axis, as there is no correlation between the two (Ruffins and Ettensohn, 1996; Summers *et al.*, 1996).

Although the evidence is inconclusive at this point, these studies are consistent with the view that the O-Ab axis is entrained before fertilization in indirect developing sea urchin embryos and that cell fates are partly specified relative to asymmetries set up along the axis. In some species, the position of the first mitotic spindle is tightly linked to the future O-Ab axis by as yet unknown mechanisms, whereas in other species the two are essentially uncoupled.

B. Blastomere Isolation/Recombination Studies

When early cleavage stage embryos are bisected to separate the prospective oral and aboral poles, the half embryos form two complete pluteus larvae, indicating that cell fates are not yet committed at this stage and that patterning along the O-Ab axis involves cell–cell interactions (Hörstadius, 1973; Cameron *et al.*, 1996). The isolated oral half typically develops with the polarity of the original O-Ab axis, but the aboral half sometimes develops with reversed polarity (Hörstadius, 1973; McCain and McClay, 1994; Henry and Raff, 1994). To explain these results, it has been suggested that a putative regulator might be present in a gradient along the O-Ab axis, such that the region of highest concentration stabilizes the oral pole, whereas the region of lower concentration exhibits unstable polarity (Hörstadius, 1973; McCain and McClay, 1994). The ability of oral and aboral halves to regulate is greatly reduced if the embryo is bisected at the late blastula stage (Hörstadius, 1973; McCain and McClay, 1994; Henry and Raff, 1994). This suggests that fates are committed along this axis by that stage. Consistent with such a view, the first indications of polarized gene expression along the O-Ab axis become apparent at the late blastula stage (see earlier discussion). In addition, the period of sensitivity of the O-Ab axis to nickel ions immediately precedes gastrulation (Hardin *et al.*, 1992). All these findings suggest that polarization of the embryo along the O-Ab axis, although entrained by maternal cues, is not rigidly fixed until the blastula stage, presumably as a result of cell–cell interactions.

Studies on the cell interactions that underlie O-Ab patterning have focused on the ectoderm, which derives largely from the eight mesomeres that constitute the animal half of the embryo. As described earlier, animal halves isolated at early cleavage stages form spherical embryoids with squamous epithelium on one side and thickened epithelium with long cilia on the other (Hörstadius, 1973; Livingston and Wilt, 1990; Wikramanayake *et al.*, 1995; Wikramanayake and Klein, 1997). Marks placed at the vegetal pole of the animal half come to lie at the boundary between the two types of epithelium, indicating that the animal half embryoids are polarized along the O-Ab axis and not the A-V axis (Hörstadius, 1973). Consistent with this view, neurons form only in the thickened region of the animal half embryoids (Wikramanayake and Klein, 1997). In *S. purpuratus*, the aboral ectoderm-specific marker *Spec1* is restricted to the squamous epithelium of the embryoids, further supporting the hypothesis that this is the aboral pole of the embryoid. In *L. pictus*, however, animal half embryoids do not express the *Spec1* homolog (*LpS1*) or another aboral ectoderm-specific marker, *LpC2* (Wikramanayake *et al.*, 1995). In both species, O-Ab polarity is abnormal in the sense that EctoV expression is not restricted to one side of the embryoid, but is instead expressed by all cells (Livingston and Wilt, 1990; Wikramanayake *et al.*, 1995). Following recombination of an animal half with micromeres or treatment with appropriate concentrations of LiCl, however, EctoV becomes restricted to one surface (Livingston and Wilt, 1990; Wikramanayake and Klein, 1997). Similar results have been found for the Hpoe antigen in *H. pulcherrimus* (Yoshikawa, 1997). Thus, the expression of certain oral ectoderm-specific features by mesomere derivatives does not require interactions with vegetal cells, but vegetal signaling is necessary for their proper restriction to one side of the embryo.

Although these studies show clearly that vegetally derived signals are required for oral–aboral polarization of the ectoderm, several important issues remain unresolved concerning (a) the timing of vegetal-to-animal signals, (b) the numbers of different signals involved, and (c) whether the signals are polarized along the O-Ab axis. With respect to the issue of timing, Henry *et al.* (1989) noted a significant increase in the ability of animal caps to form a ciliary band (a manifestation of O-Ab polarity) between the 8- and 16-cell stages, indicating that vegetal signals are transmitted at this very early stage. Wikramanayake *et al.* (1995) have also proposed that vegetal-to-animal signals might be transmitted during early cleavage and have speculated that differences in the timing of such signals might explain differences in the expression of aboral ectoderm markers in animal halves isolated from *S. purpuratus* and *L. pictus*. They suggest that vegetally derived signals may be required for the upregulation of aboral ectoderm-specific mRNAs in both species, but if they are transmitted somewhat earlier in *S. purpuratus* (i.e., during or very rapidly after the third cleavage division), then genes such as *Spec1* might be activated in animal halves after removal of vegetal blastomeres at the 8-cell stage.

Other evidence indicates, however, that signals transmitted from the vegetal hemisphere at much later developmental stages influence ectoderm development. Vegetal signals that restrict the Hpoe antigen appear to be transmitted between mesenchyme blastula and gastrula stages (Yoshikawa, 1997). The formation of the stomodeum, an oral ectoderm derivative, occurs at only a low frequency in animal halves isolated from 16-cell, 32-cell, or even mesenchyme blastula stage embryos (Henry *et al.*, 1989; Malinda and Ettensohn, 1994), but at much higher frequencies when the isolation is carried out at the early gastrula stage (Hardin and Armstrong, 1997). This suggests that vegetal-to-animal signals required for stomodeum formation are transmitted relatively late in development, between the mesenchyme blastula and early gastrula stages (Hardin and Armstrong, 1997). Presumably such signals are transmitted in a planar fashion through the ectodermal epithelium. One interpretation of these various findings is that vegetal-to-animal signals that pattern the ectoderm along the O-Ab axis are transmitted at (at least) two different developmental stages: early cleavage and late blastula.

Whether several distinct signals emanate from the vegetal region is obscure at present. The development of some oral ectoderm-specific features (EctoV expression and neuron formation) does not require the presence of vegetal cells; however, a signal is necessary to suppress the expression of these features in the prospective aboral ectoderm. Wikramanayake and Klein (1997) therefore postulate a negative, vegetally derived signal that suppresses oral ectoderm differentiation in the aboral region. They further suggest, however, that there may be a second, positive signal required to activate expression of aboral ectoderm-specific genes in that region. This two-signal model is based on the finding that the restricted expression of EctoV and Spec1 can be separated experimentally; animal half embryos from *S. purpuratus*, or animal half embryos from *L. pictus* treated with LiCl, can show polarized expression of Spec1, whereas the EctoV antigen is expressed globally (Wikramanayake and Klein, 1995, 1997). At present, it is not possible to distinguish the activities of two separate molecular signals from some difference in the timing or strength of a single signal.

Finally, a central issue is whether there is a polarized response of the animal cells to a uniform signal (e.g., polarized competence), as envisioned by Wikramanayake and Klein (1997), or whether the vegetally derived signals are themselves polarized along the O-Ab axis. In *Xenopus*, where it is possible to distinguish the polarity of early cleavage stage embryos along the secondary (D-V) axis based on morphological markers, dorsal and ventral blastomeres from the vegetal region have been shown to elicit different responses when combined with animal blastomeres (Dale and Slack, 1987). In the sea urchin, similar recombination experiments are not currently feasible due to a lack of reliable visible markers of the secondary axis in early cleavage stage embryos. In any event, it is worth noting that both the "polarized signal" and "polarized response" models require that asymmetry along the O-Ab axis be established prior to the interaction.

C. BMP2/4 Signaling

In *Drosophila* and vertebrates, patterning along the secondary (dorsal–ventral) axis is regulated by the polarized expression of BMP4 and its antagonists, Noggin and Chordin (see DeRobertis and Sasai, 1996). A BMP2/4 homolog has been cloned from two sea urchin species, *S. purpuratus* and *L. variegatus*, and shown to be ~90% identical to vertebrate BMP2/4 family members within the C-terminal portion that represents the mature ligand (Angerer *et al.*, 2000). Sea urchin BMP2/4 mRNA is expressed transiently at the blastula and early gastrula stages in a region that corresponds approximately to the prospective ectoderm. Expression is polarized orthogonal to the A-V axis, with the highest levels of mRNA on the oral side of the embryo.

Overexpression of BMP2/4 (or *Xenopus* BMP4) by microinjection of mRNA into eggs produces embryos with radialized, squamous ectoderm that expresses high levels of the Spec1 protein, generally considered to be a marker for aboral ectoderm, and only low levels of the EctoV and UH2-95 antigens, markers for oral ectoderm/ciliary band. Conversely, overexpression of *Xenopus* Noggin produces radialized ectoderm with oral-like characteristics and a radialized PMC pattern similar to those produced by treatment of embryos with nickel (Hardin *et al.*, 1992). Misregulation of BMP2/4 expression can also affect differentiation along the A-V axis; at higher doses, the gut is reduced in size and the formation of pigment cells is partially suppressed, whereas Noggin overexpression appears to increase the amount of endoderm. A metalloprotease related to *Tolloid* and BMP1 is expressed in the animal region of the very early blastula (Reynolds *et al.*, 1992; Lepage *et al.*, 1992a). Overexpression of the intact molecule has potent ventralizing effects on *Xenopus* embryos, whereas a putative dominant-negative form lacking the protease domain has a complementary, dorsalizing effect (Wardle *et al.*, 1999). Overexpression of the intact molecule in sea urchins produces a phenotype similar to that of overexpressing BMP2/4, as expected for a positive effector of BMP signaling (Angerer and Angerer, 1999).

Many questions remain concerning the role of BMP2/4 in patterning. One challenge will be to sort out the role of the pathway in polarization along the primary and secondary axes. For example, it is not yet known whether misregulation of BMP2/4 signaling alters ectodermal fates along the O-A axis directly or indirectly by perturbing vegetal-to-animal interactions. The animalized phenotype that results from overexpression of BMP2/4 is qualitatively less severe than that caused by perturbation of β-catenin signaling; for example, gut and primary mesenchyme cells still form in embryos overexpressing BMP2/4. This suggests that BMP2/4 is not involved in regulating early specification events along the A-V axis, such as micromere specification and early mesoderm/endoderm induction. This is consistent with the relatively late (blastula stage) accumulation of the BMP2/4 transcript, which provides further evidence that BMP2/4 signaling acts distinctly later than the (maternal) β-catenin pathway. Another issue is that the predominantly oral distribution of the BMP2/4 mRNA seems contradictory to the finding that overexpression of the molecule stimulates the formation of aboral-like ectoderm, whereas antagonism with Noggin promotes oral ectoderm (or at least ciliary band) formation (Angerer *et al.*, 2000). The situation would almost certainly be clarified by the isolation of additional molecular markers for ectodermal territories, especially for oral ectoderm. Obviously, it will be important to identify other regulators of BMP2/4 signaling (Noggin, Chordin, BMP2/4 receptors, and SMADs) in the sea urchin and determine the patterns of expression of these molecules, as well as that of mature BMP2/4 protein.

D. Analysis of *cis*-Regulatory Regions of Genes Differentially Expressed along the Oral–Aboral Axis

One approach that has been taken to examine patterning along the oral–aboral axis is the analysis of the *cis*-regulatory regions of genes that are asymmetrically expressed at early stages along this axis. Four genes analyzed in this regard are the two cytoskeletal actins, *CyIIIa* and *CyIIIb*, the *Spec2a* gene, which encodes a member of the calmodulin–troponin C superfamily of calcium-binding proteins, and arylsulfatase. All four genes are first expressed during cleavage and their expression is restricted to the aboral ectoderm by the late blastula stage (Cox *et al.*, 1986).

The *cis*-regulatory region of the *CyIIIa* gene, comprising ~2.3 kb of DNA upstream of the gene, contains at least 20 regulatory sites to which bind at least nine different transcription factors (Calzone *et al.*, 1988; Thézé *et al.*, 1990). It is organized into proximal, middle, and distal modules, each of which has a distinct regulatory function, although important interactions also occur between elements of the different modules (Kirchhamer and Davidson, 1996; Kirchhamer *et al.*, 1996). The proximal module appears to control the earliest expression of *CyIIIa*; it suffices to drive aboral ectoderm-specific activation of reporter genes at the time when *CyIIIa* is normally first expressed, and its activity (assayed by quantitative analysis of reporter gene expression) peaks before the other modules (Kirchhamer and Davidson, 1996). When the P1 site of this module is mutated, transcription is almost completely inactivated (Franks *et al.*, 1990; Kirchhamer and Davidson, 1996). There are additional, positive regulatory sites in the *CyIIIa* proximal module, but these seem to have a secondary role (Kirchhamer and Davidson, 1996). Mutations in the P3A site cause ectopic expression in the oral ectoderm, indicating that this site is an important negative spatial regulator.

Mutation of the *cis*-regulatory region rarely results in ectopic expression in the prospective endoderm or mesoderm of early embryos, suggesting that positive regulators of the proximal module are active only in the prospective ectoderm

(Kirchhamer and Davidson, 1996). Kirchhamer and Davidson (1996) have suggested that maternal P1 factor, the key positive regulator of the proximal module, is localized or active only in the animal-most two-thirds of the embryo by the late cleavage stage. The repressor SpP3A2 may be present in all prospective ectoderm in an inactive form but, as a consequence of early specification processes, is activated only in the prospective oral ectoderm, where it suppresses *CyIIIa* expression. Alternatively, it could initially be present in an active form and become inactivated in the prospective aboral ectoderm. This latter possibility would be consistent with the model of Wikramanayake and Klein (1997) if vegetal signals are necessary for the expression of *CyIIIa* as they are for other aboral ectoderm-specific genes. Five variants of SpP3A2 have been observed and are suggested to represent different phosphorylation derivatives that correspond to inactive and active forms of the transcription factor (Harrington *et al.*, 1992). It is not known whether the phosphorylation state of SpP3A2 is influenced by vegetal signaling.

The middle module controls *CyIIIa* expression after the blastula stage (Coffman *et al.*, 1996; Kirchhamer and Davidson, 1996). The factors that bind to this module differ from those that bind the proximal module and include the positive regulator, SpRunt (Coffman *et al.*, 1996), and two negative spatial regulators, SpZ12-1 and SpMyb (Kirchhamer and Davidson, 1996; Wang *et al.*, 1995; Coffman *et al.*, 1997). The *SpMyb* transcript has been shown to be expressed in the oral ectoderm, mesenchyme, and endoderm, consistent with the proposed role of this factor as a negative regulator (Coffman *et al.*, 1997). SpGCF1 sites present throughout the *cis*-regulatory region are thought to be involved in the looping of the DNA to allow interaction between the modules (Zeller *et al.*, 1995) and to convey positive control over the expression of *CyIIIa* (Kirchhamer and Davidson, 1996).

The 2.2-kb *cis*-regulatory region of the *CyIIIb* gene contains at least four regions of protein–DNA interactions (Niemeyer and Flytzanis, 1993). The most proximal region is nearly identical to a large part of the proximal module of *CyIIIa* (Flytzanis *et al.*, 1989) and is probably involved in the early activation of the gene. One important difference between the proximal regions of *CyIIIa* and *CyIIIb*, however, is that *CyIIIb* lacks a P3A site (Flytzanis *et al.*, 1989). The distal sites of the *CyIIIb* promoter, C1L, C1R, and E1, have no similarity to the upstream region of *CyIIIa*, but do exert negative spatial control (Flytzanis *et al.*, 1989; Niemeyer and Flytzanis, 1993). The C1L and E1 sites suppress expression in the oral ectoderm, and the C1R site suppresses expression in the endoderm and mesoderm (Xu *et al.*, 1996). The transcription factor SpCOUP-TF binds to steroid hormone response elements within the C1R site (Chan *et al.*, 1992; Niemeyer and Flytzanis, 1993). As described earlier, *SpCOUP-TF* mRNA is localized in the oocyte, egg, and early embryo at 45° relative to the O-Ab axis in two of the three species examined (Vlahou *et al.*, 1996). Its possible localization near the oral pole is consistent with the possibility that SpCOUP-TF plays a role in suppressing *CyIIIb* expression in the prospective oral ectoderm; however, this is inconsistent with the proposed function of the C1R element in suppressing expression in the endoderm and mesoderm (Xu et al., 1996). Thus the C1R site may bind to other transcription factors to result in this effect. A possible candidate is SpSHR2, another orphan steroid hormone receptor (Kontrogianni-Konstantopoulos et al., 1996; Vlahou et al., 1996).

The *cis*-regulatory region of the *Spec2a* gene includes 1.5 kb of upstream DNA, which is sufficient to convey the correct temporal and spatial activation of reporters (Gan et al., 1990; Gan and Klein, 1993; Mao et al., 1994). It is divided into three subregions: a distal region, which prevents expression in the mesenchyme and includes sites for the P3A and SpMyb transcription factors; a proximal region required for basal promoter activity; and a middle RSR element, which is necessary and sufficient for expression in the aboral ectoderm and mesenchyme (Gan and Klein, 1993; Mao et al., 1994). Within the RSR region are multiple binding sites for the Otx transcription factor (Gan and Klein, 1993; Mao et al., 1994). Mutation of the Otx-binding sites completely blocks expression, and a regulatory sequence containing two Otx-binding sites is sufficient for aboral ectoderm-specific expression (Mao et al., 1994). Four *SpOtx* transcripts encode two proteins (SpOtxα and SpOtxβ), which differ only in their N-terminal regions and the timing of their expression; *SpOtxα* is expressed first, followed by *SpOtxβ* (Li et al., 1997). Using an antibody that recognizes both forms of the protein, SpOtx was found in the cytoplasm of all cells in early cleavage stage embryos (Mao et al., 1996). Beginning at the 60- to 120-cell stage, when *Spec2a* is first transcribed only in the aboral ectoderm, the SpOtx protein is translocated to the nuclei of all cells (Gan et al., 1995; Mao et al., 1996). This is consistent with the possibility that the nuclear translocation of SpOtx provides temporal control of *Spec2a* activation, although the lack of any regional specificity in the translocation suggests that negative factors within prospective endoderm, mesoderm, and oral ectoderm must suppress activation in these cell types. The localization of SpOtx to the nuclei of nonaboral ectodermal cells also suggests that this transcription factor might play a role in the specification of other cell types. Overexpression of SpOtxα, but not SpOtxβ, causes the differentiation of aboral ectoderm cells at the expense of other cell types, supporting the hypothesis that the early form plays a role in the specification of aboral ectoderm (Mao et al., 1996; Li et al., 1997).

Otx may also be involved in activating the expression of arylsulfatase specifically in the aboral ectoderm. The first intron of the *HpArs* gene contains tandem repeats of Otx-binding sites and has positive control over transcription (Iuchi et al., 1995; Sakmoto et al., 1997). However, the late form, HpOtxL (which is homologous to SpOtxβ), but not the early form, HpotxE (which is homologous to SpOtxα), has the ability to activate the *HpArs* promoter in gastrula stage embryos (Kiyama et al., 1998).

An unexpected finding that emerges from these studies is that the key negative and positive regulators that have been identified are generally different for the different genes expressed in the aboral ectoderm. It seems unlikely that each gene expressed in this territory is controlled by a distinct mechanism; there may be common features that have not yet emerged from the analyses to date.

VII. A Model of Early Patterning

A unified scheme of the early patterning of the sea urchin embryo is presented in Fig. 4. This is an extremely fragile model, but it integrates much information and should serve to focus further work on key issues. It is similar in many respects to another analysis (Davidson *et al.*, 1998), but differs in (1) a revised view of the major early territories of gene expression, (2) a greater emphasis on autonomous specification processes within early blastomeres and less on a cascade of vegetal-to-animal signaling, and (3) a focus on β-catenin as a key regulator of autonomous specification processes and as a link between patterning along A-V and O-Ab axes. Davidson *et al.* (1998) also envisioned a number of additional cell interactions, including signals from aboral and oral polar blastomeres to lateral blastomeres, negative interactions across the endoderm–ectoderm boundary, and interactions among ectoderm cells.

The patterning of the early embryo can be viewed as the result of an interplay between autonomous, maternally based specification processes and a progression of intercellular signaling events, themselves entrained by the maternal system. A central feature of patterning is a maternally derived system that controls the differential nuclearization of β-catenin along the A-V axis. Based primarily on work with *Xenopus*, as well as initial studies with the sea urchin (Logan *et al.*, 1999), we suggest that regulation of β-catenin nuclearization along the A-V axis is controlled in a cell-autonomous fashion by molecular efforts acting independently of Wnt ligands and *frizzled* class receptors. As an example, Fig. 4 (left) shows a graded expression along the A-V axis of a maternal, positive regulator of β-catenin nuclearization (e.g., an inhibitor of GSK3). As discussed previously, there are several potential positive and negative regulators of β-catenin nuclearization that might be differentially distributed along the A-V axis. As shown in Fig. 4, the maternal, β-catenin-based system is responsible for controlling the autonomous regional expression of VEB genes, and presumably other early zygotically expressed genes, along the A-V axis.

We also propose the existence of a prefertilization O-Ab axis, based primarily on the differential localization of maternal mRNAs such as *SpCOUP* and *COLL1*α orthogonal to the A-V axis. The O-Ab axis is not rigidly fixed in the egg or early embryo; the maternal axis may be envisioned as biasing subsequent cell signaling processes that establish the O-Ab axis in a progressive manner. To reflect this distinction, we hypothesize the existence of an O-Ab "regulator," rather than a "determinant" in the strictest sense. It is unknown whether the prefertilization molecular axis is repositioned with respect to the site of sperm entry or the cytoskeletal architecture of the 1-cell zygote. In some species (but not others), a tight linkage between the position of the first cleavage plane and the O-Ab axis suggests that this may be the case. The differential localization of the putative maternal regulator along the O-Ab axis is proposed to drive the polarized expression of certain early zygotic genes orthogonal to the A-V axis, as shown in Fig. 4. Indeed, polarized expression of some zygotic genes orthogonal

1. Patterning the Early Sea Urchin Embryo

to the A-V axis takes places as early as the 16-cell stage, as evidenced by the expression of *PlHbox12*. [It should be noted, however, that some evidence suggests that the expression of this particular gene is not regulated solely by a maternal, cell autonomous system but by cell–cell interactions (DiBernardo *et al.*, 1995).]

The β-catenin pathway is a link between patterning along primary and secondary axes. With respect to the secondary polarization of the ectoderm, it is clear that vegetally derived signals play a central role. As noted earlier, it is unknown whether the patterning of the ectoderm via such cues is a consequence of polarized signals, a polarized response of the animal cells to a uniform signal, or a combination of both. Because β-catenin never accumulates at detectable levels in the nuclei of mesomeres or their descendants, we hypothesize that β-catenin influences ectodermal patterning through an effect on vegetal cells, namely by establishing a polarized, vegetally derived signal that secondarily patterns the ectoderm along the O-Ab axis, thereby activating the expression of aboral ectoderm genes in the aboral region (Fig. 4). The signaling properties of one cell type, the micromeres, have been shown to be dependent on levels of nuclear β-catenin (Logan *et al.*, 1999), and it seems likely that autonomously derived signaling properties of other vegetal blastomeres will prove to be dependent as well. We can only speculate concerning the role of β-catenin in establishing such a polarized vegetal signal. The nuclear localization of β-catenin in the vegetal region of the embryo might itself be asymmetric along the O-Ab axis (although this has not been detected), or there may be a cooperative effect in the vegetal region of the embryo between nuclear β-catenin and an asymmetrically distributed O-Ab regulator, as illustrated in Fig. 4. The hypothesis that nuclear β-catenin is required in the vegetal region for the expression of signals that pattern the ectoderm along the O-Ab axis might be tested by combining animal quartets from normal 8-cell stage embryos with cadherin (or GSK3-)-expressing vegetal quartets, and vice versa.

Figure 4 illustrates the major inductive cell interactions that occur prior to the start of gastrulation. In most cases, the timing of the interaction has not been precisely defined, but there is strong evidence that several key interactions occur during early cleavage. As discussed earlier, micromere progeny transmit signals to overlying cells between the fourth and the sixth cleavages, although they continue to be capable of transmitting signals much later in development. A critical consequence of micromere signaling is the downregulation of apical Notch protein in the central region of the vegetal plate, which is required for the specification of a subset of veg2-derived cells as nonskeletogenic mesoderm. Notch downregulation occurs at the early blastula stage, providing further evidence that critical micromere–veg2 interactions occur prior to that stage. Micromeres are also required for the normal specification of endoderm, although this appears to involve a Notch-independent mechanism.

Signals among mesomeres that suppress their ability to form mesoderm and endoderm are transmitted between the fourth and the sixth cleavages (Henry *et al.*, 1989). The timing of veg2–veg1 signaling has not been firmly established, but presumably it occurs after the formation of the veg2 layer at the 60-cell stage and prior to the establishment of the ectoderm–endoderm boundary at the mesenchyme blastula stage (see Logan and McClay, 1998, and earlier discussion). Signals from the vegetal hemisphere that polarize the ectoderm along the O-Ab axis appear to be initiated as early as the 8- to 16-cell stage, although vegetal-to-animal signals required for stomodeum formation are transmitted much later in development, namely between the late blastula and early gastrula stages (see earlier discussion). There may, in fact, be more than one vegetal-to-animal signal (Wikramanayake and Klein, 1997). In the absence of more definitive evidence, we simplify our model by showing vegetal-to-animal signals occurring at the blastula stage (Fig. 4, right).

These various intercellular signals are superimposed on basal, autonomous programs of blastomere specification. The maternal β-catenin system, presumably acting in concert with other molecular differences along the A-V axis (see Davidson *et al.*, 1998), endows early blastomeres with different regional identities and entrains autonomous specification programs within the cells. [One indication that the β-catenin system is not the *sole* specifier of regional blastomere identities, however, comes from the observation that even the most severe disruptions of β-catenin levels lead to swimming embryos with a rudimentary polarity along a single axis (Emily-Fenouil *et al.*, 1998; Wikramanayake *et al.*, 1998b; Logan *et al.*, 1999).] In some cases, it appears that autonomous and signal-dependent programs may be separated experimentally. For example, in the absence of micromeres, the veg2 progeny exhibit a modified program of mesoderm specification (Sweet *et al.*, 1999). Given that micromere-based signaling and the allocation of an early ingressing, skeletogenic mesenchyme appear to be evolutionary modifications of an ancestral echinoderm developmental program, this autonomous mesodermal specification program may reflect a more ancient plan. Similarly, it may be that the differentiative capacities of veg1 cells following veg2 removal reflect their autonomous specification program. The possibility that signals are transmitted during the brief period following a given cleavage but before blastomeres can be separated by microsurgical methods remains a lingering possibility.

The early patterning of the sea urchin embryo is similar in fundamental respects to that of *Xenopus,* the deuterostome that has probably been studied most intensively in this regard (see Harland and Gerhart, 1997; Heasman, 1997). Each system involves vegetally derived signals that are required for the formation of mesoderm and/or endoderm and for robust gastrulation movements. Each uses a maternal system of β-catenin nuclearization to establish these vegetal signaling centers. Moreover, it appears that in both systems, the vegetal signaling centers induce expression of a shared set of target genes, including *Brachyury* and *forkhead/HNF-3B* (Smith *et al.*, 1991; Ruiz i Altaba and Jessell, 1992; Harada *et al.*, 1995, 1996). In *Xenopus*, the older view that vegetally derived signals act on essentially naive blastomeres has been modified; a more current view holds that maternal factors establish initial (and labile) differences between cell groups, which are then amplified by intercellular signaling (Heaysman, 1997; Zhang *et al.*,

1998). In addition to the maternal β-catenin system, evidence shows that the BMP2/4 pathway regulates patterning along the secondary axis in both systems. One apparent difference, however, is that the Notch pathway has not been implicated in mesoderm induction in *Xenopus*, as it has been in the sea urchin. Previously, it seemed that a major difference between the two was that signaling events in *Xenopus* were thought to occur prior to the onset of zygotic transcription. More recent studies, however, have blurred this distinction as well (Zhang *et al.*, 1998). It has been proposed that another difference is that major territories of gene expression in the sea urchin embryo are more tightly coupled to early (cleavage stage) lineage compartments (Davidson, 1989, 1990). As discussed earlier, however, strong evidence of such coupling is found only with respect to micromere-derived lineages. In fact, the most significant difference between early patterning in *Xenopus* and indirect developing sea urchins may lie in the nature and extent of the cytoplasmic rearrangements that take place following fertilization.

Acknowledgments

The authors thank many members of the community of sea urchin developmental biologists for communicating experimental findings prior to publication, including K. Akasaka, L. Angerer, R. Angerer, F. Emily-Fenouil, C. Gache, W. Klein, D. Kurokawa, C. Logan, D. McClay, D. Sherwood, and A. Wikramanayake. C.A.E.'s research is supported by NSF Grant IBN-9817988, NIH Grant HD24690, a NIH Research Career Development Award, and the Stowers Institute for Medical Research.

Note added in proof: Kenny *et al.* (1999) have cloned SpSoxB1, an essential transcriptional regulator of *SpAN*, one of the VEB genes. *SpSoxB1* mRNA is distributed uniformly in the egg and early embryo, but at the 16-cell stage the protein is concentrated in the nuclei primarily of macromeres and mesomeres. *SpSoxB1* is transcribed zygotically in a "non-vegetal" pattern that reflects the corresponding domain of *SpAN* mRNA accumulation at the early blastula stage. Angerer and Angerer (2000) note that the early polarity in nuclear accumulation of SpSoxB1 protein, coupled with the complementary pattern of nuclear β-catenin localization, means that each of the three major cell types (micromeres, macromeres, and mesomeres) of the 16-cell stage embryo contains a unique complement of transcriptional regulators.

References

Aberle, H., Bauer, A., Stappert, J., Kispert, A., and Kemler, R. (1997). β-Catenin is a target for the ubiquitin-proteasome pathway. *EMBO J.* **16,** 3797–3804.
Angerer, L. M., and Angerer, R. C. (1999). Regulative development of the sea urchin embryo: Signaling cascades and morphogen gradients. *Semin. Cell Dev. Biol.* **10,** 327–334.
Angerer, L. M., and Angerer, R. C. (2000). Animal–vegetal patterning mechanisms in the early sea urchin embryo. *Dev. Biol.* **218,** 1–12.
Angerer, L. M., Dolecki, G. J., Gagnon, M. L., Lum, R., Wang, G., Yang, Q., Humphreys, T., and Angerer, R. C. (1989). Progressively restricted expression of a homeobox gene within the aboral ectoderm of developing sea urchin embryos. *Genes Dev.* **3,** 370–383.
Angerer, L. M., Oleksyn, D., Logan, C. Y., McClay, D. R., Dale, L., and Angerer, R. C. (2000). A BMP pathway regulates cell fate allocation along the sea urchin animal–vegetal embryonic axis. *Development* **127,** 1105–1114.

Angerer, R. C., and Angerer, L. M. (1997). Fate specification along the sea urchin embryo animal-vegetal axis. *Biol. Bull.* **192,** 175–177.

Angerer, R. C., Reynolds, S. D., Grimwade, J., Hurley, D. L., Yang, Q., Kingsley, P. D,. Gagnon, M. L., Palis, J., and Angerer, L. M. (1990). Contributions of the spatial analysis of gene expression to the study of sea urchin development. *In* "In Situ Hybridisation: Application to Developmental Biology" (N. Harris and D. G. Williams, eds.), pp. 69–95. Cambridge Univ. Press, Cambridge, UK.

Artavanis-Tsakonas, S., Rand, M. D., and Lake, R. J. (1999). Notch signaling: Cell fate control and signal integration in development. *Science* **284,** 770–776.

Benink, H., Wray, G., and Hardin, J. (1997). Archenteron precursor cells can organize secondary axial structures in the sea urchin embryo. *Development* **124,** 3461–3470.

Benson, S. C., Sucov, H. M., Stephens, L., Davidson, E. H., and Wilt, F. H. (1987). A lineage-specific gene encoding a major matrix protein of the sea urchin embryo spicule. I. Authentication of the cloned gene and its developmental expression. *Dev. Biol.* **120,** 499–506.

Burke, R. D. (1989). Echinoderm metamorphosis: comparative aspects of the change in form. *In* "Echinoderm Studies" (M. Jangoux and J. Lawrence, eds.), pp. 81–108. A. A. Balkema Pub., Rotterdam.

Calzone, F. J., Hoog, C., Teplow, D. B., Cutting, A. E., Zeller, R. W., Britten, R. J., and Davidson, E. H. (1991). Gene regulatory factors of the sea urchin embryo. I. Purification by affinity chromatography and cloning of P3A2, a novel DNA binding protein. *Development* **112,** 335–350.

Calzone, F. J., Thézé, N., Thiebaud, P., Hill, R. L., Britten, R. J., and Davidson, E. H. (1988). Developmental appearance of factors that bind specifically to cis-regulatory sequences of a gene expressed in the sea urchin embryo. *Genes Dev.* **2,** 1074–1088.

Cameron, R. A., Fraser, S. E., Britten, R. J., and Davidson, E. H. (1989). The oral-aboral axis of a sea urchin embryo is specified by first cleavage. *Development* **106,** 641–647.

Cameron, R. A., Britten, R. J., and Davidson, E. H. (1993). The embryonic ciliated band of the sea urchin, *Strongylocentrotus purpuratus,* derives from both oral and aboral ectoderm territories. *Dev. Biol.* **160,** 369–376.

Cameron, R. A., Leahy, P. S., and Davidson, E. H. (1996). Twins raised from separated blastomeres develop into sexually mature *Strongylocentrotus purpuratus. Dev. Biol.* **178,** 514–519.

Chan, S.-M., Xu, N., Niemeyer, C. C., Bone, J. R., and Flytzanis, C. N. (1992). SpCOUP-TF: A sea urchin member of the steroid/thyroid hormone receptor family. *Proc. Natl. Acad. Sci. USA* **89,** 10568–10572.

Chen, S. W., and Wessel, G. M. (1996). Endoderm differentiation *in vitro* identifies a transitional period for endoderm ontogeny in the sea urchin embryo. *Dev. Biol.* **175,** 57–65.

Coffman, J. A., Kirchhamer, C. V., Harrington, M. G., and Davidson, E. H. (1996). SpRunt-1, a new member of the runt-domain family of transcription factors, is a positive regulator of the aboral ectoderm-specific CyIIIa gene in sea urchin embryos. *Dev. Biol.* **174,** 43–54.

Coffman, J. A., Kirchhamer, C. V., Harrington, M. G., and Davidson, E. H. (1997). SpMyb functions as an intramodular repressor to regulate spatial expression of CyIIIa in sea urchin embryos. *Development* **124,** 4717–4727.

Coffman, J. A., and McClay, D. R. (1990). A hyaline layer protein that becomes localized to the oral ectoderm and foregut of sea urchin embryos. *Dev. Biol.* **140,** 93–104.

Cox, K. H., Angerer, L. M., Lee, J. J., Davidson, E. H., and Angerer, R. C. (1986). Cell lineage-specific programs of expression of multiple actin genes during sea urchin embryogenesis. *J. Mol. Biol.* **188,** 159–172.

Czihak, G. (1971). Echinoids. *In* "Experimental Embryology of Marine and Freshwater Invertebrates" (G. Reverberi, ed.), pp. 363–506. North-Holland, Amsterdam.

Dale, L., and Slack, J. M. W. (1987). Regional specificity within the mesoderm of early embryos of *Xenopus laevis. Development* **100,** 279–295.

Dan, K., and Tanaka, Y. (1990). Attachment of one spindle pole to the cortex in unequal cleavage. *Ann. N.Y. Acad. Sci.* **582,** 108–119.

Davidson, E. H. (1986). "Gene Activity in Early Development" 3rd ed., Academic Press, New York.

Davidson, E. H. (1989). Lineage-specific gene expression and the regulative capacities of the sea urchin embryo. *Development* **105**, 421–445.
Davidson, E. H., Cameron, R. A., and Ransick, A. (1998). Specification of cell fate in the sea urchin embryo: Summary and some proposed mechanisms. *Development* **125**, 3269–3290.
DeRobertis, E. M., and Sasai, Y. (1996). A common plan for dorsoventral patterning in Bilateria. *Nature* **380**, 27–40.
DiBernardo, M., Castagnetti, S., Bellomonte, D., Oliveri, P., Melfi, R., Palla, F., and Spinelli, G. (1999). Spatially restricted expression of P10tp, a *Paracentrotus lividus* orthopedia-related homeobox gene, is correlated with oral ectodermal patterning and skeletal morphogenesis in late-cleavage stage embryos. *Development* **126**, 2171–2179.
DiBernardo, M., Russo, R., Oliveri, P., Melfi, R., and Spinelli, G. (1995). Homeobox-containing gene transiently expressed in a spatially restricted pattern in the early sea urchin embryo. *Proc. Natl. Acad. Sci. USA* **92**, 8180–8184.
DiCarlo, M., Romancino, D. P., Montana, G., and Ghersi, G. (1994). Spatial distribution of two maternal messengers in *Paracentrotus lividus* during oogenesis and embryogenesis. *Proc. Natl. Acad. Sci. USA* **91**, 5622–5626.
Duncan, L., Alper, S., Arigoni, F., Losick, R., and Stragier, P. (1995). Activation of cell-specific transcription by a serine phosphatase at the site of asymmetric division. *Science* **270**, 641–644.
Emily-Fenouil, F., Ghiglione, C,. Lhomond, G., Lepage, T., and Gache, C. (1998). GSK3β/shaggy mediates patterning along the animal-vegetal axis of the sea urchin embryo. *Development* **125**, 2489–2498.
Ettensohn, C. A. (1992). Cell interactions and mesodermal cell fates in the sea urchin embryo. *Development (Suppl.)* 43–51.
Ettensohn, C. A. (1999). Cell movements in the sea urchin embryo. *Curr. Opin. Gen. Dev.* **9**, 461–465.
Ettensohn, C. A., Guss, K. A., Hodor, P. G., and Malinda, K. M. (1997). The morphogenesis of the skeletal system of the sea urchin embryo. *In* "Reproductive Biology of Invertebrates" (J. R. Collier, ed.). Oxford and IBM Publishing Co., New Delhi.
Ettensohn, C. A., and McClay, D. R. (1988). Cell lineage conversion in the sea urchin embryo. *Dev. Biol.* **125**, 396–409.
Eyal-Giladi, H. (1997). Establishment of the axis in chordates: Facts and speculations. *Development* **124**, 2285–2296.
Flytzanis, C. N., Bogosian, E. A., and Niemeyer, C. C. (1989). Expression and structure of the CyIIIb actin gene of the sea urchin *Strongylocentrotus purpuratus*. *Mol. Reprod. Dev.* **1**, 208–218.
Franks, R. R., Anderson, R., Moore, F. G., Hough-Evans, B. R., Britten, R. J., and Davidson, E. H. (1990). Competitive titration in living sea urchin embryos of regulatory factors required for expression of the CyIIIa actin gene. *Development* **110**, 31–40.
Gagnon, M. L., Angerer, L. M., and Angerer, R. C. (1992). Posttranscriptional regulation of ectoderm-specific gene expression in early sea urchin embryos. *Development* **114**, 457–467.
Gambino, R., Romancino, D. P., Cervello, M., Vizzini, A., Isola, M. G., Virruso, L., and Di Carlo, M. (1997). Spatial distribution of collagen type I mRNA in *Paracentrotus lividus* eggs and embryos. *Biochem. Biophys. Res. Commun.* **238**, 334–337.
Gan, L., and Klein, W. H. (1993). A positive cis-regulatory element with a bicoid target site lies within the sea urchin Spec2a enhancer. *Dev. Biol.* **157**, 119–132.
Gan, L., Mao, C.-A., Wikramanayake, A,. Angerer, L. M., Angerer, R. C., and Klein, W. H. (1995). An orthodenticle-related protein from *Srongylocentrotus purpuratus*. *Dev. Biol.* **167**, 517–528.
Gan, L., Wessel, G. M., and Klein, W. (1990). Regulatory elements from the related Spec genes of *Strongylocentrotus purpuratus* yield different spatial patterns with a lacZ reporter gene. *Dev. Biol.* **142**, 346–359.
Gan, L., Zhang, W., and Klein, W. (1990). Repetitive DNA sequences linked to the sea urchin Spec genes contain transcriptional enhancer-like elements. *Dev. Biol.* **139**, 166–196.
Ghiglione, C., Emily-Fenouil, F., Chang, P., and Gache, C. (1996). Early gene expression along the animal-vegetal axis in sea urchin embryoids and grafted embryos. *Development* **122**, 3067–3074.

Ghiglione, C., Emily-Fenouil, F., Lhomond, G., and Gache, C. (1997). Organization of the proximal promoter of the hatching-enzyme gene, the earliest zygotic gene expressed in the sea urchin embryo. *Eur. J. Biochem.* **250,** 502–513.

Ghiglione, C., Lhomond, G., Lepage, T., and Gache, C. (1993). Cell-autonomous expression and position-dependent repression by Li^+ of two zygotic genes during sea urchin early development. *EMBO J.* **12,** 87–96.

Gibson, A. W., and Burke, R. D. (1985). The origin of pigment cells in embryos of the sea urchin *Strongylocentrotus purpuratus. Dev. Biol.* **107,** 414–419.

Goldstein, B., and Freeman, G. (1997). Axis specification in animal development. *BioEssays* **19,** 105–116.

Grimwade, J. E., Gagnon, M. L., Yang, Q., Angerer, R. C., and Angerer, L. M. (1991). Expression of two mRNAs encoding EGF-related proteins identifies subregions of sea urchin embryonic ectoderm. *Dev. Biol.* **143,** 44–57.

Guss, K. A., and Ettensohn, C. A. (1997). Skeletal morphogenesis in the sea urchin embryo: Regulation of primary mesenchyme gene expression and skeletal rod growth by ectoderm-derived cues. *Development* **124,** 1899–1908.

Han, M. (1997). Gut reaction to Wnt signaling in worms. *Cell* **90,** 581–584.

Harada, Y., Akasaka, K., Shimada, H., Peterson, K. J., Davidson, E. H., and Satoh, N. (1996). Spatial expression of a forkhead homologue in the sea urchin embryo. *Mech. Dev.* **60,** 163–173.

Harada, Y., Yasuo, H., and Satoh, N. (1995). A sea urchin homologue of the chordate *Brachyury (T)* gene is expressed in the secondary mesenchyme founder cells. *Development* **121,** 2747–2754.

Hardin, J., and Armstrong, N. (1997). Short-range cell-cell signals control ectodermal patterning in the oral region of the sea urchin embryo. *Dev. Biol.* **182,** 134–149.

Hardin, J., Coffman, J. A., Black, S. D., and McClay, D. R. (1992). Commitment along the dorsoventral axis of the sea urchin embryo is altered in response to $NiCl_2$. *Development* **116,** 671–685.

Hardin, P. E., Angerer, L. M., Hardin, S. H., Angerer, R. C., and Klein, W. H. (1988). The Spec2 genes of *Strongylocentrotus purpuratus*: Structure and differential expression in embryonic aboral ectoderm cells. *J. Mol. Biol.* **202,** 417–431.

Harkey, M. A., Klueg, K., Sheppard, P., and Raff, R. A. (1995). Structure, expression, and extracellular targetting of PM27, a skeletal protein associated specifically with growth of the sea urchin larval spicule. *Dev. Biol.* **168,** 549–566.

Harkey, M. A., Whiteley, H. R., and Whiteley, A. H. (1988). Coordinate accumulation of five transcripts in the primary mesenchyme during skeletogenesis in the sea urchin embryo. *Dev. Biol.* **125,** 381–395.

Harkey, M. A., Whiteley, H. R., and Whiteley, A. H. (1992). Differential expression of the msp130 gene among skeletal lineage cells in the sea urchin embryo: A three dimensional in situ hybridization analysis. *Mech. Dev.* **37,** 173–184.

Harland, R., and Gerhart, J. (1997). Formation and function of Spemann's organizer. *Annu. Rev. Cell Dev. Biol.* **13,** 611–667.

Harrington, M. G., Coffman, J. A., Calzone, F. J., Hood, L. E., Britten, R. J., and Davidson, E. H. (1992). Complexity of sea urchin nuclear proteins that contain basic domains. *Proc. Natl. Acad. Sci. USA* **89,** 6252–6256.

Hart, M. J., de los Santos, R., Albert, I. N., Rubinfeld, B., and Polakis, P. (1998). Downregulation of β-catenin by human Axin and its association with the APC tumor suppressor, β-catenin and GSK3β. *Curr. Biol.* **8,** 573–581.

Heasman, J. (1997). Patterning the *Xenopus* blastula. *Development* **124,** 4179–4191.

Henry, J. J. (1998). The development of dorsoventral and bilateral axial properties in sea urchin embryos. *Semin. Cell Dev. Biol.* **9,** 43–52.

Henry, J. J., Amemiya, S., Wray, G. A., and Raff, R. A. (1989). Early inductive interactions are involved in restricting cell fates of mesomeres in sea urchin embryos. *Dev. Biol.* **136,** 140–153.

Henry, J. J., Klueg, K. M., and Raff, R. A. (1992). Evolutionary dissociation between cleavage, cell lineage and embryonic axes in sea urchin embryos. *Development* **114,** 931–938.

1. Patterning the Early Sea Urchin Embryo

Henry, J. J., and Raff, R. A. (1990). Evolutionary change in the process of dorsoventral axis determination in the direct developing sea urchin, *Heliocidaris erythrogramma*. *Dev. Biol.* **141**, 55–69.

Henry, J. J., and Raff, R. A. (1994). Progressive determination of cell fate along the dorsoventral axis in the sea urchin *Heliocidaris erythrogramma*. *Roux's Arch. Dev. Biol.* **204**, 62–69.

Holy, J., and Schatten, G. (1991). Differential behavior of centrosomes in unequally dividing blastomeres during fourth cleavage of sea urchin embryos. *J. Cell Sci.* **98**, 423–431.

Hoppler, S., Brown, J. D., and Moon, R. T. (1996). Expression of a dominant-negative Wnt blocks induction of MyoD in *Xenopus* embryos. *Genes Dev.* **10**, 2805–2817.

Hörstadius, S. (1973). "Experimental Embryology of Echinoderms." Clarendon Press, Oxford.

Hough-Evans, B. R., Franks, R. R., Cameron, R. A., Britten, R. J., and Davidson, E. H. (1987). Correct cell type-specific expression of a fusion gene injected into sea urchin eggs. *Dev. Biol.* **121**, 576–579.

Hough-Evans, B. R., Franks, R. R., Zeller, R. W., Britten, R. J., and Davidson, E. H. (1990). Negative spatial regulation of the lineage-specific CyIIIa actin gene in the sea urchin embryo. *Development* **110**, 41–50.

Ikeda, S., Kishida, S., Yamamoto, H., Murai, H., Koyama, S., and Kikuchi, A. (1998). Axin, a negative regulator of the Wnt signaling pathway, forms a complex with GSK-3β and β-catenin and promotes GSK-3β-dependent phosphorylation of β-catenin. *EMBO J.* **17**, 1371–1384.

Iuchi, Y., Morokuma, J., Akasaka, K., and Shimada, H. (1995). Detection and characterization of the cis-element in the first intron of the Ars gene in the sea urchin. *Dev. Growth Differ.* **37**, 373–378.

Jan, Y. N., and Jan, L. Y. (1998). Asymmetric cell division. *Nature* **392**, 775–778.

Jeffery, W. R. (1992). Axis determination in sea urchin embryos: From confusion to evolution. *Trends Genet.* **8**, 223–225.

Kenny, A. P., Kozlowski, D. J., Oleksyn, D. W., Angerer, L. M., and Angerer, R. C. (1999). SpSoxB1, a maternally encoded transcription factor asymmetrically distributed among early sea urchin blastomeres. *Development* **126**, 5473–5483.

Khaner, O., and Wilt, F. (1991). Interactions of different vegetal cells with mesomeres during early stages of sea urchin development. *Development* **112**, 881–890.

Killian, C. E., and Wilt, F. H. (1989). The accumulation and translation of a spicule matrix protein mRNA during sea urchin development. *Dev. Biol.* **133**, 148–156.

Kingsley, P. D., Angerer, L. M., and Angerer, R. C. (1993). Major temporal and spatial patterns of gene expression during differentiation of the sea urchin embryo. *Dev. Biol.* **155**, 216–234.

Kirchhamer, C. V., and Davidson, E. H. (1996). Spatial and temporal information processing in the sea urchin embryo: Modular and intramodular organization of the CyIIIa gene cis-regulatory system. *Development* **122**, 333–348.

Kirchhamer, C. V., Yuh, C.-H., and Davidson, E. H. (1996). Modular cis-regulatory organization of developmentally expressed genes: Two genes transcribed territorially in the sea urchin embryo, and additional examples. *Proc. Natl. Acad. Sci. USA* **93**, 9322–9328.

Kissinger, J. C., and Raff, R. A. (1998). Evolutionary changes in sites and timing of actin gene expression in embryos of the direct- and indirect-developing sea urchins, *Heliocidaris erythrogramma* and *H. tuberculata*. *Dev. Genes Evol.* **208**, 82–93.

Kitajima, T., and Okazaki, K. (1980). Spicule formation *in vitro* by the descendants of precocious micromeres formed at the 8-cell stage of sea urchin embryos. *Dev. Growth Differ.* **22**, 265–279.

Kiyama, T., Akasaka, K., Takata, K., Mitsunaga-Nakatsubo, K., Sakamoto, N., and Shimada, H. (1998). Structure and function of a sea urchin orthodenticle-related gene (Hpotx). *Dev. Biol.* **193**, 139–145.

Kiyomoto, M., and Shirai, H. (1993). The determinant for archenteron formation in starfish: Co-culture of an animal egg fragment-derived cell cluster and a selected blastomere. *Dev. Growth Differ.* **35**, 99–105.

Klein, P. S., and Melton, D. A. (1996). A molecular mechanism for the effect of lithium on development. *Proc. Natl. Acad. Sci. USA* **93**, 8455–8459.

Klueg, K. M., Harkey, M. A., and Raff, R. A. (1997). Mechanisms of evolutionary changes in timing, spatial expression, and mRNA processing in the msp130 gene in a direct-developing sea urchin, *Heliocidaris erythrogramma*. *Dev. Biol.* **182**, 121–133.

Kominami, T. (1988). Determination of the dorsoventral axis in early embryos of the sea urchin, *Hemicentrotus pulcherrimus*. *Dev. Biol.* **127**, 187–196.

Kontrogianni-Konstantopoulos, A., Vlahou, A., Vu, D., and Flytzanis, C. N. (1996). A novel sea urchin nuclear receptor encoded by alternatively spliced maternal RNAs. *Dev. Biol.* **177**, 371–382.

Kozlowski, D. J., Gagnon, M. L., Marchant, J. K., Reynolds, S. D., Angerer, L. M., and Angerer, R. C. (1996). Characterization of a *SpAN* promoter sufficient to mediate correct spatial regulation along the animal-vegetal axis of the sea urchin embryo. *Dev. Biol.* **176**, 95–107.

Krufka, A., Johnson, R. G., Wylie, C. C., and Heasman, J. (1998). Evidence that dorsal-ventral differences in gap junctional communication in the early *Xenopus* embryo are generated by β-catenin independent of cell adhesion effects. *Dev. Biol.* **200**, 92–102.

Kuraishi, R., and Osanai, K. (1994). Contribution of maternal factors and cellular interaction to determination of archenteron in the starfish embryo. *Development* **120**, 2619–2628.

Kurokawa, D., Kitajima, T., Mitsunaga-Nakatsubo, K., Amemiya, S., Shimada, H., and Akasaka, K. (1999). *Hp-ets*, an ets-related transcription factor implicated in primary mesenchyme cell differentiation in the sea urchin embryo. *Mech. Dev.* **80**, 41–52.

Laney, J. D., and Hochstrasser, M. (1999). Substrate targeting in the ubiquitin system. *Cell* **97**, 427–430.

Langelan, R. E., and Whiteley, A. H. (1985). Unequal cleavage and the differentiation of echinoid primary mesenchyme. *Dev. Biol.* **109**, 464–475.

Larabell, C. A., Torres, M., Rowning, B. A., Yost, C., Miller, J. R., Wu, M., Kimelman, D., and Moon, R. T. (1997). Establishment of the dorso-ventral axis in *Xenopus* embryos is presaged by early asymmetries in β-catenin that are modulated by the Wnt signaling pathway. *J. Cell. Biol.* **136**, 1123–1136.

Lee, J. J., Calzone, F. J., Britten, R. J., and Davidson, E. H. (1986). Activation of sea urchin actin genes during embryogenesis: Measurement of transcript accumulation from five different genes in *Strongylocentrotus purpuratus*. *J. Mol. Biol.* **188**, 173–183.

Lepage, T., Ghiglione, C., and Gache, C. (1992a). Spatial and temporal expression pattern during sea urchin embryogenesis of a gene coding for a protease homologous to the human protein BMP-1 and to the product of the *Drosophila* dorsal-ventral patterning gene tolloid. *Development* **114**, 147–164.

Lepage, T., Sardet, C., and Gache, C. (1992b). Spatial expression of the hatching enzyme gene in the sea urchin embryo. *Dev. Biol.* **150**, 23–32.

Leyns, L., Bouwmeister, T., Kim, S.-H., Piccolo, S., and DeRobertis, E. M. (1997). Frz-b is a secreted antagonist of Wnt signaling expressed in the Spemann organizer. *Cell* **88**, 747–756.

Li, N., Chuang, C.-K., Mao, C.-A., Angerer, L. M,. and Klein, W. H. (1997). Two Otx proteins generated from multiple transcripts of a single gene in *Strongylocentrotus purpuratus*. *Dev. Biol.* **187**, 253–266.

Livant, D. L., Cutting, A., Britten, R. J., and Davidson, E. H. (1988). An in vivo titration of regulatory factors required for expression of a fusion gene in transgenic sea urchin embryos. *Proc. Natl. Acad. Sci. USA* **85**, 7607–7611.

Livingston, B. T., and Wilt, F. H. (1989). Lithium evokes expression of vegetal-specific molecules in the animal blastomeres of sea urchin embryos. *Proc. Natl. Acad. Sci. USA* **86**, 3669–3673.

Livingston, B. T., and Wilt, F. H. (1990). Range and stability of cell fate determination in isolated sea urchin blastomeres. *Development* **108**, 403–410.

Logan, C. Y., and McClay, D. R. (1997). The allocation of early blastomeres to the ectoderm and endoderm is variable in the sea urchin embryo. *Development* **124**, 2213–2223.

Logan, C. Y., and McClay, D. R. (1998). The lineages that give rise to the endoderm and mesoderm in the sea urchin embryo. In "Cell Fate and Lineage Determination" (S. Moody, ed.), pp. 41–58. Academic Press, New York.

Logan, C. Y., Miller, J. R., Ferkowicz, M. J., and McClay, D. R. (1999). Nuclear β-catenin is required to specify vegetal cell fates in the sea urchin embryo. *Development* **126**, 345–357.

1. Patterning the Early Sea Urchin Embryo

Luke, N. H., Killian, C. E., and Livingston, B. T. (1997). *Spfkh1* encodes a transcription factor implicated in gut formation during sea urchin development. *Dev. Growth Differ.* **39**, 285–294.

Malinda, K. M., and Ettensohn, C. A. (1994). Primary mesenchyme cell migration in the sea urchin embryo: Distribution of directional cues. *Dev. Biol.* **164**, 562–578.

Mao, C.-A., Gan, L., and Klein, W. H. (1994). Multiple Otx binding sites required for expression of the *Strongylocentrotus purpuratus* Spec2a gene. *Dev. Biol.* **165**, 229–242.

Mao, C.-A., Wikramanayake, A. H., Gan, L., Chaung, C.-K., Summers, R. G., and Klein, W. H. (1996). Altering cell fates in sea urchin embryos by overexpressing SpOtx, an orthodenticle-related protein. *Development* **122**, 1489–1498.

Maruyama, Y. K., Nakaseko, Y., and Yagi, S. (1985). Localization of cytoplasmic determinants responsible for primary mesenchyme formation and gastrulation in the unfertilized egg of the sea urchin *Hemicentrotus pulcherrimus*. *J. Exp. Zool.* **236**, 155–163.

McCain, E. R., and McClay, D. R. (1994). The establishment of bilateral asymmetry in sea urchin embryos. *Development* **120**, 395–404.

McClay, D. R., and Logan, C. Y. (1996). Regulative capacity of the archenteron during gastrulation in the sea urchin. *Development* **122**, 607–616.

Miller, J. R., and McClay, D. R. (1997). Changes in the pattern of adherens junction-associated β-catenin accompany morphogenesis in the sea urchin embryo. *Dev. Biol.* **192**, 310–322.

Miller, J. R., and Moon, R. T. (1996). Signal transduction through β-catenin and specification of cell fate during embryogenesis. *Genes Dev.* **10**, 2527–2539.

Miller, R. N., Dalamagas, D. G., Kingsley, P. D., and Ettensohn, C. A. (1996). Expression of *S9* and actin *CyIIa* mRNAs reveals dorso-ventral polarity and mesodermal sublineages in the vegetal plate of the sea urchin embryo. *Mech. Dev.* **60**, 3–12.

Minokawa, T., and Amemiya, S. (1999). Timing of the potential of micromere-descendants in echinoid embryos to induce endoderm differentiation of mesomere descendants. *Develop. Growth Differ.* **41**, 535–547.

Minokawa, T., Hamaguchi, Y., and Amemiya, S. (1997). Skeletogenic potential of induced secondary mesenchyme cells derived from the presumptive ectoderm in echinoid embryos. *Dev. Genes Evol.* **206**, 472–476.

Molenaar, M., van de Wetering, M., Oosterwegel, M. Peterson-Maduro, J., Godsave, S., Korinek, V., Roose, J., Destree, O., and Clevers, H. (1996). XTcf-3 transcription factor mediates beta-catenin-induced axis formation in *Xenopus* embryos. *Cell* **86**, 391–399.

Montana, G., Romancino, D. P., and diCarlo, M. D. (1996). Cloning, expression, and localization of a new member of a *Paracentrotus lividus* cell surface multigene family. *Mol. Reprod. Dev.* **44**, 36–43.

Moon, R. T., Brown, J. D., and Torres, M. (1997). WNTs modulate cell fate and behavior during vertebrate development. *Trends Genet.* **13**, 157–162.

Morisato, D., and Anderson, K. V. (1995). Signaling pathways that establish the dorsal-ventral pattern of the *Drosophila* embryo. *Annu. Rev. Genet.* **29**, 371–399.

Morris, V. B. (1995). Apluteal development of the sea urchin *Holopneustes purpurescens Agassiz* (Echinodermata: Echinoidea: Euechinoidea). *Zool. J. Linn. Soc.* **114**, 349–364.

Niemeyer, C. C., and Flytzanis, C. N. (1993). Upstream elements involved in the embryonic regulation of the sea urchin CyIIIb actin gene: Temporal and spatial specific interactions at a single cis-acting element. *Dev. Biol.* **156**, 293–302.

Okazaki, K. (1975). Spicule formation by isolated micromeres of the sea urchin embryo. *Am. Zool.* **15**, 567–581.

Okazaki, K., Fukushi, T., and Dan, D. (1962). Cyto-embryological studies of sea urchins. IV. Correlation between the shape of the ectodermal cells and the arrangement of the primary mesenchyme cells in sea urchin larvae. *Acta Embryol. Morphol. Exp.* **5**, 17–31.

Olson, D. J., Christian, J. L., and Moon, R. T. (1991). Effect of Wnt-1 and related proteins on gap junctional communication in *Xenopus* embryos. *Science* **252**, 1173–1176.

Orford, K., Crockett, C., Jensen, J. P., Weissman, A. M., and Byers, S. W. (1997). Serine phosphorylation-regulated ubiquitination and degradation of β-catenin. *J. Biol. Chem.* **272**, 24735–24738.

Pehrson, J. R., and Cohen, L. H. (1986). The fate of the small micromeres in sea urchin development. *Dev. Biol.* **113**, 522–526.
Qi, H., Rand, M. D., Wu, X., Sestan, N., Wang, W., Rakic, P., Xu, T., and Artavanis-Tsakonas, S. (1999). Processing of the Notch ligand Delta by the metalloprotease Kuzbanian. *Science* **283**, 91–94.
Raff, E. C., Popodi, E. M., Sly, B. J., Turner, F. R., Villinski, J. T., and Raff, R. A. (1999). A novel ontogenetic pathway in hybrid embryos between species with different modes of development. *Development* **126**, 1937–1945.
Raff, R. A. (1992). Direct-developing sea urchins and the evolutionary reorganization of early development. *Bioessays* **14**, 211–218.
Ransick, A., and Davidson, E. H. (1993). A complete second gut induced by transplanted micromeres in the sea urchin embryo. *Science* **259**, 1134–1138.
Ransick, A., and Davidson, E. H. (1995). Micromeres are required for normal vegetal plate specification in sea urchin embryos. *Development* **121**, 3215–3222.
Ransick, A., and Davidson, E. H. (1998). Late specification of veg1 lineages to endodermal fate in the sea urchin embryo. *Dev. Biol.* **195**, 38–48.
Ransick, A., Ernst, S., Britten, R. J., and Davidson, E. H. (1993). Whole mount *in situ* hybridization shows *Endo16* to be a marker for the vegetal plate territory in sea urchin embryos. *Mech. Dev.* **42**, 117–124.
Reynolds, S. D., Angerer, L. M., Palis, J., Nasir, A., and Angerer, R. C. (1992). Early mRNAs, spatially restricted along the animal-vegetal axis of sea urchin embryos, include one encoding a protein related to tolloid and BMP-1. *Development* **114**, 769–786.
Roberson, M., Neri, A., and Oppenheimer, S. B. (1975). Distribution of concanavalin A receptor sites on specific populations of embryonic cells. *Science* **189**, 639–640.
Roose, J., Molenaar, M., Peterson, J., Hurnkamp, J., Brantjes, H., Moerer, P., van de Wetering, M., Destree, O., and Clevers, H. (1998). The *Xenopus* Wnt effector XTcf-3 interacts with Groucho-related transcriptional repressors. *Nature* **8**, 608–612.
Ruffins, S. W., and Ettensohn, C. A. (1996). A fate map of the vegetal plate of the sea urchin (*Lytechinus variegatus*) mesenchyme blastula. *Development* **122**, 253–263.
Ruiz i Altaba, A., and Jessell, T. M. (1992). Pintallavis, a gene expressed in the organizer and midline cells of frog embryos: involvement in the development of the neural axis. *Development* **116**, 81–93.
Sakamoto, N., Akasaka, K., Nitsunaga-Nakatsubo, K., Takata, K., Nishitani, T., and Shimada, H. (1997). Two isoforms of orthodenticle-related proteins (Hpotx) bind to the enhancer element of sea urchin arylsulfatase gene. *Dev. Biol.* **181**, 284–295.
Schatten, G. (1981). Sperm incorporation, the pronuclear migrations, and their relation to the establishment of the first embryonic axis: Time lapse video microscopy of the movements during fertilization of the sea urchin *Lytechinus variegatus*. *Dev. Biol.* **86**, 426–437.
Schnabel, R., and Priess, J. R. (1997). "*C. elegans* II," pp. 361–382. CSHL Press, New York.
Schroeder, T. E. (1980a). Expression of the prefertilization polar axis in sea urchin eggs. *Dev. Biol.* **79**, 428–443.
Schroeder, T. E. (1980b). The jelly canal: Marker of polarity for sea urchin oocytes, eggs, and embryos. *Exp. Cell Res.* **128**, 490–494.
Schroeder, T. E. (1981). Development of a "primitive" sea urchin (*Eucidaris tribuloides*): Irregularities in the hyaline layer, micromeres, and primary mesenchyme. *Biol. Bull.* **161**, 141–151.
Seeling, J. M., Miller, J. R., Gil, R., Moon, R. T., White, R., and Virshup, D. M. (1999). Regulation of β-catenin signaling by the B56 subunit of protein phosphatase 2A. *Science* **283**, 2089–2091.
Senger, D. R., and Gross, P. R. (1978). Macromolecule synthesis and determination in sea urchin blastomeres at the sixteen-cell stage. *Dev. Biol.* **65**, 404–415.
Sharrocks, A. D., Brown, A. L., Ling, Y., and Yates, P. R. (1997). The ETS-domain transcription factor family. *Int. J. Biochem. Cell Biol.* **29**, 1371–1387.
Sherwood, D. R., and McClay, D. R. (1997). Identification and localization of a sea urchin Notch ho-

1. Patterning the Early Sea Urchin Embryo   43

mologue: Insights into vegetal plate regionalization and Notch receptor regulation. *Development* **124,** 3363–3374.
Sherwood, D. R., and McClay, D. R. (1999). LvNotch signaling mediates secondary mesenchyme specification in the sea urchin embryo. *Development* **126,** 1703–1713.
Slack, J. M. (1991). "From Egg to Embryo: Regional Specification in Early Development." Cambridge Univ. Press, Cambridge.
Smith, J. C., Price, B. M., Green, J. B. A., Weigel, D., and Merrmann, B. G. (1991). Expression of a *Xenopus* homolog of Brachyury (T) is an immediate–early response to mesoderm induction. *Cell* **67,** 79–87.
Sokol, S. Y. (1996). Analysis of dishevelled signalling pathways during *Xenopus* development. *Curr. Biol.* **6,** 1456–1467.
Stambolic, V., Ruel, L., and Woodgett, J. R. (1996). Lithium inhibits glycogen synthase kinase-3 activity and mimics Wingless signalling in intact cells. *Curr. Biol.* **6,** 1664–1668.
Summers, R., Morrill, J., Leith, A., Marko, M., Piston, D., and Stonebrakes, A. (1903). A stereometric analysis of karyokinesis, cytokinesis and cell arrangements during and following fourth cleavage period in the sea urchin, *Lytechinus variegatus*. *Dev. Growth Diff.* **35,** 41–57.
Summers, R. G., Piston, D. W., Harris, K. M., and Morrill, J. B. (1996). The orientation of first cleavage in the sea urchin embryo, *Lytechinus variegatus*, does not specify the axes of bilateral symmetry. *Dev. Biol.* **175,** 177–183.
Sweet, H., Hodor, P. G., and Ettensohn, C. A. (1999). The role of micromere signaling in Notch activation and mesoderm specification during sea urchin embryogenesis. *Development* **126,** 5255–5265.
Tanaka, Y. (1976). Effects of surfactants on the cleavage and further development of sea urchin embryos. 1. The inhibition of micromere formation at the fourth cleavage. *Dev. Growth* and *Differ.* **18,** 113–122.
Tanaka, Y. (1981). Distribution and redistribution of pigment granules in the development of sea urchin embryos. *Wilh. Roux Arch.* **190,** 267–273.
Thézé, N., Calzone, F. J., Thiebaud, P., Hill, R. L., Britten, R. J., and Davidson, E. H. (1990). Sequences of the CyIIIa actin gene regulatory domain bound specifically by sea urchin embryo nuclear proteins. *Mol. Reprod. Dev.* **25,** 110–122.
van Weeren, P. C., de Bruyn, K. M., de Vries-Smits, A. M., van Lint, J., and Burgering, B. M. (1998). Essential role for protein kinase B (PKB) in insulin-induced glycogen synthase kinase 3 inactivation. *J. Biol. Chem.* **273,** 13150–13156.
Vlahou, A., Gonzalez-Rimbau, M., and Flytzanis, C. N. (1996). Maternal mRNA encoding the orphan steroid receptor SpCOUP is localized in sea urchin eggs. *Development* **122,** 521–526.
Vonica, A., Weng, W., Gumbiner, B. M., and Venuti, J. M. (2000). TCF is the nuclear effector of the beta-catenin signal that patterns the sea urchin animal–vegetal axis. *Dev. Biol.* **217,** 230–243.
Waltzer, L., and Bienz, M. (1998). *Drosophila* CBP represses the transcription factor TCF to antagonize Wingless signalling. *Nature* **395,** 521–525.
Wang, D. G.-W., Kirchhamer, C. V., Britten, F. J., and Davidson, E. H. (1995). SpZ12-1, a negative regulator required for spatial control of the territory-specific CyIIIa gene in the sea urchin embryo. *Development* **121,** 1111–1122.
Wang, S., Krinks, M., Lin, K., Luyten, F. P., Moos, M., (1997). Frzb, a secreted protein expressed in the Spemann organizer, binds and inhibits wnt-8. *Cell* **88,** 757–766.
Wang, W., Wikramanayake, A. H., Gonzalez-Rimbau, M., Vlahou, A., Flytzanis, C. N., and Klein, W. H. (1996). Very early and transient vegetal plate expression of *SpKrox1*, a *Krüppel/Krox* gene from *Strongylocentrotus purpuratus*. *Mech. Dev.* **60,** 185–195.
Wardle, F., Angerer, L. M., Angerer, R. C., and Dale, L. (1999). Regulation of BMP signaling by the MP1/TLD-related metalloprotease, SpAN. *Dev. Biol.* **206,** 63–72.
Wasylyk, B., Hagman, J., and Gutierrez-Hartmann, A. (1998). Ets transcription factors: Nuclear effectors of the Ras-MAP-kinase signaling pathway. *Trends Biochem. Sci.* **23,** 213–216.

Wei, Z., Angerer, L. M,. and Angerer, R. C. (1997a). Multiple positive *cis*-elements regulate the asymmetric expression of the *SpHE* gene along the sea urchin embryo animal-vegetal axis. *Dev. Biol.* **187,** 71–78.

Wei, Z., Angerer, L. M., and Angerer, R. C. (1999a). Spatially regulated SpEts4 transcription factor activity along the sea urchin embryo animal-vegetal axis. *Development* **126,** 1729–1737.

Wei, Z., Angerer, R. C., and Angerer, L. M. (1999b). Identification of a new sea urchin Ets protein, SpEts4, by yeast one-hybrid screening with the hatching enzyme promoter. *Mol. Cell. Biol.* **19,** 1271–1278.

Wei, Z., Angerer, L. M., Gagnon, M. L., and Angerer, R. C. (1995). Characterization of the SpHE promoter that is spatially regulated along the animal-vegetal axis of the sea urchin embryo. *Dev. Biol.* **171,** 195–211.

Wei, Z., Kenny, A. P., Angerer, L. M., and Angerer, R. C. (1997b). The *SpHE* gene is downregulated in sea urchin late blastulae despite persistance of multiple positive factors sufficient to activate its promoter. *Mech. Dev.* **67,** 171–178.

Wessel, G. M., Goldberg, L., Lennarz, W. J., and Klein, W. H. (1989). Gastrulation in the sea urchin is accompanied by the accumulation of an endoderm-specific mRNA. *Dev. Biol.* **136,** 526–536.

Wessel, G. M., and McClay, D. R. (1985). Sequential expression of germ layer specific molecules in the sea urchin embryo. *Dev. Biol.* **111,** 451–463.

Wikramanayake, A. H., Brandhorst, B. P., and Klein, W. H. (1995). Autonomous and non-autonomous differentiation of ectoderm in different sea urchin species. *Development* **121,** 1497–1505.

Wikramanayake, A. H., Huang, L., Dayal, S., and Klein, W. H. (1998a). Wnt signaling is required for gastrulation and aboral ectoderm formation in the sea urchin embryo. *Dev. Biol.* **198,** 182. [Abstract]

Wikramanayake, A. H., Huang, L., and Klein, W. H. (1998b). β-catenin is essential for patterning the maternally specified animal-vegetal axis in the sea urchin embryo. *Proc. Natl. Acad. Sci. USA* **95,** 9343–9348.

Wikramanayake, A. H., and Klein, W. H. (1997). Multiple signaling events specify ectoderm and pattern the oral-aboral axis in the sea urchin embryo. *Development* **124,** 13–20.

Wray, G. A., and McClay, D. R. (1989). Molecular heterochronies and heterotopies in early echinoid development. *Evolution* **43,** 803–813.

Wray, G. A., and Raff, R. A. (1990). Novel origins of lineage founder cells in the direct-developing sea urchin *Heliocidaris erythrogramma*. *Dev. Biol.* **141,** 41–54.

Xu, N., Niemeyer, C. C., Gonzalez-Rimbau, M., Bogosian, E. A., and Flytzanis, C. N. (1996). Distal cis-acting elements restrict expression of the CyIIIb actin gene in the aboral ectoderm of the sea urchin embryo. *Mech. Dev.* **60,** 151–162.

Yang, Q., Angerer, L. M., and Angerer, R. C. (1989). Structure and tissue-specific developmental expression of a sea urchin arylsulfatase gene. *Dev. Biol.* **135,** 53–65.

Yoshikawa, S. (1997). Oral/aboral ectoderm differentiation of the sea urchin embryo depends on a planar or secretory signal from the vegetal hemisphere. *Dev. Growth Differ.* **39,** 319–327.

Yost, C., Farr, G. H., III, Pierce, S. B., Ferkey, D. M,. Chen, M. M., and Kimelman, D. (1998). GBP, an inhibitor of GSK-3, is implicated in *Xenopus* development and oncogenesis. *Cell* **93,** 1031–1041.

Zeller, R. W., Griffith, J. D., Moore, J. G., Kirchhamer, C. V., Britten, R. J., and Davidson, E. H. (1995). A multimerizing transcription factor of sea urchin embryos capable of looping DNA. *Proc. Natl. Acad. Sci. USA* **92,** 2989–2993.

Zeng, L., Fagotto, F., Zhang, T., Hsu, W., Vasicek, T. J., Perry, W. L., III, Lee, J. J., Tilghman, S. M., Gumbiner, B. M., and Costantini, F. (1997). The mouse *fused* locus encodes axin, an inhibitor of the Wnt signaling pathway that regulates embryonic axis formation. *Cell* **90,** 181–192.

Zhang, J., Houston, D. W., King, M. L., Payne, C., Wylie, C., and Heasman, J. (1998). The role of maternal VegT in establishing the primary germ layers in *Xenopus* embryos. *Cell* **94,** 515–524.

2
Turning Mesoderm into Blood: The Formation of Hematopoietic Stem Cells during Embryogenesis

Alan J. Davidson and Leonard I. Zon
Division of Hematology/Oncology
Harvard Medical School and Howard Hughes Medical Institute
Children's Hospital
Boston, Massachusetts 02115

I. Introduction
II. Mesoderm Induction and Patterning
III. Fate Maps
IV. Ventralizing Homeobox Genes
   A. The Mix Family
   B. The Vent Family
V. Specification of Hematopoietic Stem Cells
   A. The Role of SCL
   B. The Role of *cloche*
VI. Conclusions
VII. Future Directions
   References

The formation of hematopoietic stem cells during development occurs by a multistep process that begins with the induction of ventral mesoderm. This mesoderm is patterned during gastrulation by a bone morphogenetic protein (BMP) signaling pathway that is mediated, at least in part, by members of the Mix and Vent families of homeobox transcription factors. Following gastrulation, a subset of ventral mesoderm is specified to become hematopoietic stem cells. Key determinants of hematopoietic fate include the product of the zebrafish *cloche* gene and the basic helix–loop–helix transcription factor SCL. Future studies in *Xenopus* and zebrafish should reveal other critical factors in this developmental pathway. © 2000 Academic Press.

I. Introduction

Hematopoietic stem cells provide a continuous supply of mature blood cells during embryogenesis and throughout adult life. Much of our understanding of lineage commitment, differentiation, and growth control of blood cells is derived from the study of hematopoiesis in the adult (D'Andrea, 1994). Comparatively less is known about the origin and formation of hematopoietic stem cells during embryonic development. Insights into the genetic program responsible for early blood cell formation have come from the study of hematopoiesis in *Xenopus* and zebrafish embryos (Mead *et al.*, 1996; Ransom *et al.*, 1996; Gering *et al.*, 1998; Liao *et al.*, 1998; Mead *et al.*, 1998a; Thompson *et al.*, 1998). During early vertebrate development, mesoderm is induced and then patterned to give rise to the differentiated cell types found along the dorsal–ventral axis of the embryo (Sive, 1993; Heasman, 1997). Following gastrulation, a population of ventral mesoderm cells is specified as hematopoietic stem cells. Therefore, a comprehension of early blood development requires an understanding of the inductive and patterning events that give rise to ventral mesoderm and specify hematopoietic stem fate. This review focuses on recent discoveries made in *Xenopus* and zebrafish that have provided new insights into the molecular mechanisms underlying early blood cell development.

II. Mesoderm Induction and Patterning

In zebrafish and frogs, mesoderm arises from the equatorial region of the blastula embryo known as the marginal zone (for reviews, see Sive, 1993; Driever, 1995; Heasman, 1997; Solnica-Krezel, 1999). Signaling molecules released from the underlying vegetal blastomeres in *Xenopus laevis*, or the yolk cell syncytial layer in *Danio rerio*, are thought to induce the neighboring cells of the marginal zone to adopt a mesodermal fate. Early amphibian studies demonstrated that vegetal blastomeres possess two inducing capacities: induction of ventral mesoderm (prospective blood and mesenchyme) and induction of dorsal mesoderm (prospective notochord and somites). While the identities of these signals are still unclear, candidate mesoderm-inducing factors include members of the transforming growth factor-β (TGF-β) and fibroblast growth factor (FGF) families (reviewed by Smith, 1995). More recent studies have highlighted the importance of the T-box transcription factor VegT as a crucial regulator of germ layer formation. Depletion of maternal VegT transcripts inhibits the capacity of vegetal cells to form endoderm and release mesoderm-inducing signals (Zhang *et al.*, 1998). The zebrafish gene *spadetail*, which appears to be the homologue of *VegT*, is not expressed maternally but has an expression pattern similar to the zygotic expression of *VegT*. Mutant *spadetail* embryos display a severe deficit in mesodermal and endodermal derivatives in the trunk, including an absence of differentiated blood cells (Kimmel *et al.*, 1989; Ho and Kane, 1990; Thompson *et al.*, 1998).

During gastrulation, the marginal zone becomes patterned to give rise to differentiated cell types such as notochord, somites, pronephros, and blood. Coordinating these patterning events is the Spemann organizer, a small region of dorsal mesoderm that has the ability to induce a secondary axis when implanted into the ventral region of a host embryo (Spemann and Mangold, 1924). Based on early *Xenopus* fate maps it was believed that the circumference of the marginal zone represented the dorsal–ventral axis of the mesoderm (Dale and Slack, 1987a,b; Smith, 1989). Notochord, the most dorsal mesodermal derivative, was found to be derived from the dorsal marginal zone, the region of the Spemann organizer, whereas somites, pronephros, mesenchyme, and blood develop sequentially from progressively more lateral regions of the marginal zone.

Mesodermal patterning is thought to arise from an antagonistic relationship between growth factors, which promote ventral fates (ventralizing signals), and molecules secreted from the Spemann organizer, which induce dorsal fates (for reviews, see Graff, 1997; Harland and Gerhart, 1997; Thomsen, 1997). Considerable evidence suggests that the ventralizing signal is mediated by members of the bone morphogenetic protein (BMP) subgroup belonging to the TGF-β superfamily of growth factors. Overexpression of *bmp2, bmp4,* or *bmp7* during development perturbs normal mesodermal patterning by causing an expansion of ventral mesodermal fates at the expense of more dorsal derivatives (Dale *et al.*, 1992; Jones *et al.*, 1992; Fainsod *et al.*, 1994; Clement *et al.*, 1995; Wang *et al.*, 1997). Conversely, overexpression of a dominant-negative BMP receptor or antisense *bmp4* RNA inhibits ventral fates and causes an expansion of dorsal mesodermal derivatives (Graff *et al.*, 1994; Suzuki *et al.*, 1994; Sasai *et al.*, 1995; Steinbeisser *et al.*, 1995). In zebrafish, the dorsalized mutant *swirl* lacks the ventral and ventrolateral tissues blood and pronephros, respectively, and shows a concomitant expansion of dorsal and dorsolateral mesodermal derivatives (Mullins *et al.*, 1996; Nguyen *et al.*, 1998). The phenotype of *swirl* is caused by a mutation in *bmp2b*, one of the two paralogs of *bmp2* that exist in the zebrafish genome (Kishimoto *et al.*, 1997; Martinez-Barbera *et al.*, 1997; Nguyen *et al.*, 1998). Similarly, the dorsalized phenotype of *snailhouse* has recently been shown to be caused by a mutation in *bmp7* (Schmid *et al.*, 2000; Dick *et al.*, 2000). Another dorsalized mutant known as *somitobun* is caused by a defect in *smad5*, which encodes a molecule involved in mediating BMP signaling (Whitman, 1998; Hild *et al.*, 1999). Smad5 is required between the midblastula to early gastrula stages to transduce BMP signaling and to maintain *bmp2b* expression via an autoregulatory feedback loop (Hild *et al.*, 1999). In *Xenopus*, a number of dorsalizing molecules expressed in the Spemann organizer have been identified and include the secreted factors chordin, noggin, cerberus, and follistatin (Piccolo *et al.*, 1996; Zimmerman *et al.*, 1996; Iemura *et al.*, 1998; Piccolo *et al.*, 1999). Overexpression of these factors causes dorsalization of ventral mesoderm by interacting with BMPs and preventing receptor activation. Genetic evidence supports a role for chordin in mesoderm patterning. The zebrafish *chordino* mutant results from a null mutation in the zebrafish *chordin* homologue and is characterized by a reduction in anterior structures and excessive blood cell formation (Fisher *et al.*, 1997; Schulte-Merker *et al.*, 1997). The antagonism observed between Spemann organizer-derived molecules such as chordin and ventralizing factors such as BMPs has led to the proposal that a gradient of BMP signaling across the marginal zone is responsible for specifying different mesodermal fates. (Dosch *et al.*, 1997; Graff, 1997; Neave *et al.*, 1997; Thomsen, 1997; Jones and Smith, 1998). According to this model, high levels of BMP signaling specify hematopoietic cell fate in the ventral marginal zone (mesoderm furthest away from the Spemann organizer), intermediate BMP signaling specifies more lateral derivatives (pronephros and heart), whereas low levels of BMP signaling determine more dorsal fates such as somitic mesoderm.

III. Fate Maps

Previously constructed fate maps for *Xenopus* terminated before the ventral blood island (VBI) and posterior somites had completely formed (Moody, 1987a,b; Dale and Slack, 1987a,b). A more recent fate mapping study has examined the lineage of hematopoietic cells from the 32-cell stage (stage 6) to the tadpole (stage 41) when the entire VBI and all 45-plus pairs of somites are present (Lane and Smith, 1999). Blood was not restricted to the progeny of the C4 blastomere as previously thought (Dale and Slack, 1987b) but was instead derived from all vegetal hemisphere blastomeres (C and D tiers) and occasionally from animal blastomeres A3 and A4. Additional mapping at the 64-cell stage refined the blood domain to the lower and upper daughters of the C and D tiers, respectively. Based on these results, it was concluded that the blood fate map domain occupies the vegetal portion of the marginal zone at the 32-cell stage (Fig. 1).

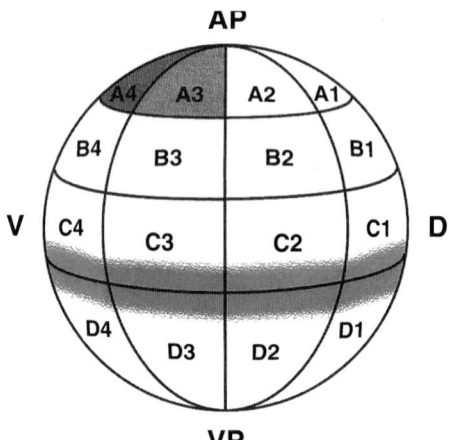

Fig. 1 *Xenopus* summary fate map for blood (32-cell stage to tadpole). Blood was found to arise from blastomeres C1–C4 and D1–D4 with a variable contribution from blastomeres A3 and A4 (shaded in gray). Refined mapping at the 64-cell stage determined that lower C- and upper D-tier progeny contribute to hematopoietic cells, thus placing the blood fate map domain in the lower portion of the marginal zone (projected onto the stage 6 embryo in gray). AP, animal pole; VP, vegetal pole; D, dorsal; V, ventral. (Adapted with permission from Lane and Smith, 1999, courtesy of Company of Biologists Ltd.)

2. Development of Hematopoietic Stem Cells   49

The progeny of the C1 blastomere are known to contribute to the Spemann organizer (Vodicka and Gerhart, 1995). The finding that this blastomere also gives rise to blood was unexpected given that the dorsal marginal zone is not considered to possess hematopoietic potential (Maeno et al., 1992; Kelley et al., 1994). Further evidence to support a dorsal contribution to blood was provided by the observation that embryos dorsalized by lithium or deuterium oxide treatment express globin (Lane and Smith, 1999). A recent expression analysis of the *Xenopus* homologue of *AML1* (encoding a runt domain transcription factor) also suggests that the VBI is populated by dorsally derived cells (Tracey et al., 1998). However, the hematopoietic identity of the dorsal cells expressing *Xaml1* was not confirmed, thus raising the possibility that this population may represent vascular cells instead (Cleaver et al., 1997; North et al., 1999).

Due to technical limitations in lineage labeling cells beyond the 64-cell stage, it was only possible to confirm that the blood territory comes within approximately 22.5° of the dorsal midline (Lane and Smith, 1999). Thus, it is not known if the blood domain occupies all 360° of the vegetal marginal zone at the 64-cell stage or whether it is excluded from the dorsal-most region. The contribution of the early gastrula marginal zone to blood has been examined by transplantation experiments between cytogenetically distinct embryos (Turpen et al., 1997). Ventral, lateral, or dorsal marginal zones were transplanted from diploid donors into triploid hosts and the contribution of these grafts to primitive blood was examined. Dorsal and lateral marginal zones failed to contribute to hematopoietic cells in the VBI. In contrast, the ventral marginal zone was found to contribute significantly to primitive blood (Turpen et al., 1997). These results indicate that despite the finding that blood descends from dorsal and ventral blastomeres at stage 6, by the early gastrula stage all prospective hematopoietic tissue is restricted to the ventral marginal zone.

IV. Ventralizing Homeobox Genes

A number of putative ventralizing transcription factors have been identified in *Xenopus* and include the Mix and Vent families of homeobox-containing genes.

A. The Mix Family

The Mix family of transcription factors belongs to the paired class of homeobox genes that is currently composed of seven members known as *Mix.1*, *Mix.2*, *Mix.3/Mixer*, *Bix1/Mix.4*, *Bix2/milk*, *Bix3*, and *Bix4* (Rosa, 1989; Vize, 1996; Ecochard et al., 198; Henry and Melton, 1998; Mead et al., 1989b; Tada et al., 1998). Transcripts for the Mix genes are found in presumptive mesoderm and/or endoderm throughout the equatorial region of the embryo. Expression of these homeobox genes is transient, starting soon after the midblastula transition (MBT) when zygotic gene transcription occurs and terminating at the end of gastrulation.

Mix.1 is the founding member of the family and the most extensively studied. Although originally isolated as an immediate early response gene to activin induction (Rosa, 1989), *Mix.1* was later reidentified in a screen for ventralizing factors (Mead *et al.*, 1996). Overexpression of *Mix.1* during development causes a loss of anterior structures and excessive blood formation, which is similar to the ventralized phenotype arising from ectopic BMP expression (Dale *et al.*, 1992; Jones *et al.*, 1992; Hemmati-Brivanlou and Thomsen, 1995). BMP-4 was found to upregulate *Mix.1* expression in animal cap explants, and a dominant-negative form of Mix.1 was shown to partially block BPM-4-induced ventralization (Mead *et al.*, 1996). These results have led to the suggestion that Mix.1 may participate in a BMP signaling pathway involved in the patterning of ventral mesoderm. Other Mix homeoproteins have been reported to ventralize, including Mix.2, Mix.3/Mixer, and Bix1/Mix.4, suggesting that additional family members may also be involved in the development of ventral fates (Mead *et al.*, 1998b; Tada *et al.*, 1998).

The Mix homeoproteins have been implicated in endoderm formation. In the case of Bix1/Mix.4, endoderm and ventral mesoderm are induced in a dose-dependent manner. Overexpression of a high level of *Bix1/Mix.4* in animal cap explants induces endoderm-specific markers, whereas a lower dose induces globin expression (Tada *et al.*, 1998). *Mix.1* cannot induce endoderm when overexpressed in animal caps (Lemaire *et al.*, 1998; Mead *et al.*, 1998b); however, when coexpressed with the related homeobox genes *goosecoid* or *siamois*, it appears to act synergistically to upregulate transcripts of endoderm markers (Lemaire *et al.*, 1998; Latinkic and Smith, 1999). In contrast, the induction of endoderm by *Mix.3/Mixer* is progressively inhibited by coinjection of increasing doses of *Mix.1* (Mead *et al.*, 1998b). These synergistic and antagonistic interactions probably arise from the ability of the Mix homeoproteins to form homo- and heterodimers with other paired-class transcription factors (Wilson *et al.*, 1993; Mead *et al.*, 1996, 1998b).

Recent studies have employed the use of "antimorphic" constructs composed of Mix.1 fused to the engrailed repressor domain to provide further insights into the function of Mix.1. The conclusions from these studies conflict with earlier observations (Mead *et al.*, 1996) and suggest that Mix.1 is required for endoderm formation rather than ventral patterning (Lemaire *et al.*, 1998; Latinkic and Smith, 1999). The ability of Mix.1 to form heterodimers with other transcription factors, coupled with the fact that multiple Mix family members are coexpressed in mesoderm and endoderm, makes it difficult to interpret loss-of-function phenotypes arising from Mix.1–repressor fusions. A more meaningful assessment of Mix function may come from a genetic approach such as that afforded by the mouse and zebrafish systems.

B. The Vent Family

Xvent-1 (which is identical or similar to *PV.1* and *Xvent-1B*) and *Xvent-2* (which is identical or similar to *Xom, Vox, Xbr-1,* and *Xvent-2B*), constitute a second group of putative ventralizing homeobox-containing genes (Gawantka *et al.*, 1995; Ladher *et al.*, 1996; Onichtchouk *et al.*, 1996; Papalopulu and Kintner, 1996; Schmidt *et al.*, 1996; Tidman Ault *et al.*, 1996). In the early gastrula, transcripts for *Xvent-1* are localized to ventral and ventrolateral regions of the marginal zone and become progressively restricted to posterior–ventral regions by the neurula stage (Gawantka *et al.*, 1995). *Xvent-2* transcripts are initially found throughout the late blastula embryo but become excluded from the organizer, neural plate, and future dorsal axis during gastrulation (Onichtchouk *et al.*, 1996; Schmidt *et al.*, 1996). The phenotype arising from *Xvent-1* or *Xvent-2* overexpression is characterized by a truncation of head structures and axial defects, which is similar in appearance to that of embryos ventralized by ectopic BMP expression (Dale *et al.*, 1992; Jones *et al.*, 1992; Hemmati-Brivanlou and Thomsen, 1995). BMP-4 can upregulate *Xvent-2* transcripts in the presence of cycloheximide, indicating that this gene is a direct target of BMP signaling and therefore a likely mediator of BMP function (Ladher *et al.*, 1996; Friedle *et al.*, 1998). Consistent with this notion, many of the dorsal genes that are downregulated by BMP signaling are similarly affected by *Xvent-1* or *Xvent-2* overexpression (Gawantka *et al.*, 1995; Schmidt *et al.*, 1996; Friedle *et al.*, 1998; Melby *et al.*, 1999). Furthermore, both Xvent-1 and Xvent-2 are able to rescue secondary axes induced by the expression of a dominant-negative BMP receptor (Onichtchouk *et al.*, 1996; Tidman Ault *et al.*, 1996). Taken together, these findings have led to the assumption that Xvent-1 and Xvent-2 participate in a BMP signaling pathway involved in the maintenance and patterning of ventral mesoderm. However, the effects of *Xvent-1/-2* overexpression on blood cell development have only recently been addressed. Surprisingly, embryos injected with *Xvent-1, Xvent-2,* or both together display decreased globin expression at stage 30/31 despite displaying a morphology similar to ventralized embryos (Kumano *et al.*, 1999; Xu *et al.*, 1999). When assayed at stage 37/38, the level of globin expression in embryos overexpressing *Xvent-1* or *Xvent-2* had increased but was still lower than that in control embryos (Kumano *et al.*, 1999). In these embryos, the anterior VBI was severely reduced or absent, consistent with anterior truncations reported in previous studies (Gawantka *et al.*, 1995; Ladher *et al.*, 1996; Onichtchouk *et al.*, 1996; Schmidt *et al.*, 1996; Tidman Ault *et al.*, 1996). These results are puzzling in light of the data implicating *Xvent-1* and *Xvent-2* as downstream effectors of BMP signaling. Further insight into the role of the Vent family of homeoproteins during blood cell development will come from the identification of a zebrafish mutant or by gene-targeting analysis in the mouse.

V. Specification of Hematopoietic Stem Cells

A. The Role of SCL

In contrast to the early formation of ventral mesoderm, the first appearance of blood cells does not occur until relatively late in development. Expression of the markers *scl, lmo2, gata-1*, and *gata-2* in hematopoietic cells is not detected by *in situ* hybridization until gastrulation has finished (Kelley *et al.*, 1994; Gering *et al.*, 1998; Liao *et al.*, 1998; Mead *et al.*, 1998a; Thompson *et al.*, 1998). The *scl* gene encodes a basic helix–loop–helix transcription factor that is expressed in hematopoietic, vascular, and neuronal tissues (reviewed by Robb and Begley, 1997). The importance of *scl* for the specification of hematopoietic stem cells has been demonstrated by gene-targeting experiments in the mouse (Porcher *et al.*, 1996; Robb *et al.*, 1996) and overexpression studies in *Xenopus* (Mead *et al.*, 1998a) and zebrafish (Gering *et al.*, 1998; Liao *et al.*, 1998). In *Xenopus, scl* is first detected at stage 15 in a small patch of cells in the ventral-most region of the embryo and, as development proceeds, this expression expands laterally and caudally, forming a V-shaped pattern characteristic of the developing VBI (Kelley *et al.*, 1994; Mead *et al.*, 1998a). In zebrafish, *scl* expression begins around the 2- to 3-somite stage in bilateral stripes of cells on either side of the embryo (Gering *et al.*, 1998; Liao *et al.*, 1998).

Expression of *scl* in bFGF-treated *Xenopus* animal pole explants causes an up-regulation of globin transcripts, suggesting that SCL is capable of activating the blood program in mesoderm (Mead *et al.*, 1998a). Ectopic expression of *scl* in dorsal marginal zone explants results in the development of blood cells, but unlike the overexpression of *bmp-4,* the formation of dorsal anterior structures such as the eye and cement gland is not inhibited (Mead *et al.*, 1998a). Thus, it has been proposed that SCL acts as a "master regulator" that activates genes controlling the specification of hematopoietic stem cells from mesoderm (Green, 1996; Gering *et al.*, 1998; Liao *et al.*, 1998; Mead *et al.*, 1998a).

Although BMPs are required for blood cell formation, a direct link between the BMP signaling cascade and the induction of *scl* has yet to be shown. A recent study in *Xenopus* has implicated the ventral trunk epidermis as a source of BMPs important for blood cell differentiation. The expression of BMP antagonists in the descendants of the A4 blastomere, which contribute to the epidermis of the trunk (Dale and Slack, 1987a), resulted in a severe reduction in globin expression in the VBI (Kumano *et al.*, 1999). Because the expression of early hematopoietic markers such as *scl* was not examined in this study, it is not clear if signaling by epidermally derived BMPs is required for hematopoietic stem cell specification or merely promotes hemoglobinization. The fact that globin transcripts can still be detected in cultured explants of ventral mesoderm in which the ectoderm has been removed suggests that epidermal signals are not required for the formation of hematopoietic stem cells (Maeno *et al.*, 1994). Instead,

BMPs secreted from the ventral ectoderm may promote erythropoiesis in a fashion similar to the role of stromal cells in the bone marrow (Mayani *et al.*, 1992; Maeno *et al.*, 1996).

B. The Role of *cloche*

The *cloche* mutant is characterized by a severe deficit in hematopoietic and endothelial cells despite the otherwise normal development of other tissues (Stainier *et al.*, 1995). Recent cell transplantation studies have addressed the cell autonomy of these defects (Parker and Stainier, 1999). With regard to vascular development, *cloche* was found to be required cell-autonomously for the differentiation of endothelial precursors. With regard to hematopoietic development, a non-cell-autonomous requirement for *cloche* was found prior to *gata-1* expression, whereas a cell-autonomous requirement was found subsequent to *gata-1* expression. Transcripts for *scl* are absent in *cloche* mutants during early segmentation stages, and expression at later stages is restricted to a small number of cells in the tail (Liao *et al.*, 1998). The lack of early *scl* expression in this mutant suggests that the *cloche* defect acts upstream of *scl* to block the specification of hematopoietic stem cells and angioblasts from ventral mesoderm. Consistent with this notion, the forced expression of *scl* in *cloche* mutants leads to a rescue of the blood and vascular defects (Liao *et al.*, 1998). The rescued blood cells in *scl*-injected embryos were found to express *gata-1* and possessed complexed hemoglobin, thus indicating that erythroid differentiation had occurred in the *cloche* background. These results suggest a model in which *cloche* acts in a cell- and non-cell-autonomous fashion during the differentiation of ventral mesoderm to blood.

VI. Conclusions

The formation of hematopoietic stem cells during embryogenesis can be considered to occur by a multistep process that involves (1) the formation of ventral mesoderm, (2) the patterning of ventral mesoderm during gastrulation, and (3) the induction of genes that specify hematopoietic stem cell fate (Fig. 2). Classical and more modern studies in amphibians have demonstrated that ventral mesoderm induction occurs in response to signals released from the ventral vegetal blastomeres and by the activity of critical transcription factors such as VegT. A ventral requirement for the zygotic expression of *VegT* is highlighted by the *spadetail* mutant, which displays a severe reduction in blood cells. BMP signaling is necessary to maintain and pattern ventral mesoderm during gastrulation. Recent studies have begun to unravel the molecular components of the BMP signaling cascade, and important downstream mediators include the Mix and

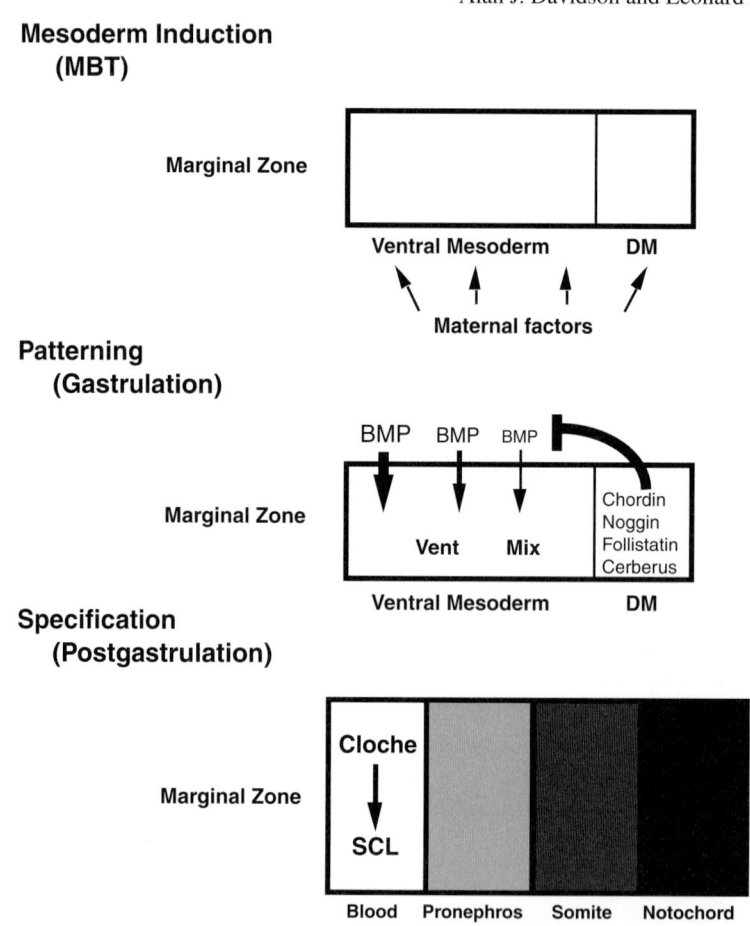

Fig. 2 Hypothetical scheme of steps leading to the formation of hematopoietic stem cells during development. For simplicity, only the marginal zone is shown (represented by a thicked-line rectangle). (1) Mesoderm is induced in the marginal zone of blastula stage embryos by maternally derived factors such as the transcription factor VegT and secreted growth factor(s) belonging to the TGF-β and FGF families. (2) During gastrulation the marginal zone is patterned by a gradient of bone morphogenetic protein (BMP) signaling, which is established by the secretion of BMP antagonists (chordin, noggin, follistatin, and cerberus) from dorsal mesoderm (DM; also known as the Spemann organizer). Important mediators of ventral mesoderm patterning include the transcription factors belonging to the Vent and Mix families of homeodomain proteins. (3) Following gastrulation, the *cloche* gene product induces the expression of *scl* (encoding a basic helix-loop-helix transcription factor) in a subset of ventral mesoderm, and hematopoietic stem cell fate is specified.

Vent families of homeobox transcription factors. Following the extensive cell movements of gastrulation, a subset of ventral mesoderm is specified to become hematopoietic stem cells. Molecular and genetic studies have identified the basic helix–loop–helix transcription factor SCL as a key determinant of hematopoietic stem cell fate. Although the molecular pathways involved in *scl* induction are unknown, it appears likely that the function of the *cloche* gene product is required to induce and maintain *scl* expression in hematopoietic cells (Liao *et al.*, 1998; Parker and Stainier, 1999).

VII. Future Directions

The revised *Xenopus* fate map for blood has revealed that blood cells descend from a much more extensive region of the stage 6 embryo than previously thought. The widespread distribution of prospective ventral mesoderm raises a number of interesting questions that need to be addressed in future studies. First, how does ventral mesoderm arise in dorsal regions of the marginal zone when current evidence suggests that dorsovegetal blastomeres only induce dorsal mesoderm? Second, how is the identity of ventral mesoderm maintained in the dorsal marginal zone in the presence of BMP antagonists? The answers to these questions will come from the further study of the pathways responsible for ventral mesoderm induction and patterning.

The molecular program governing the development of ventral fates is likely to be complex, and although the BMP pathway plays a dominant role, it is clear that positive and negative regulatory loops exist with other signaling pathways. For example, *Xvent-1* expression is positively regulated by *Wnt-8* expression and vice versa, whereas several members of the Mix family are induced by BMP-4, activin, and Vg1 (Hoppler and Moon, 1998; Mead *et al.*, 1998b). Recent data have suggested a link between VegT and the Mix family members Bix1/Mix.4 and Bix4 (Tada *et al.*, 1998; Casey *et al.*, 1999), whereas other studies have revealed cross talk between the BMP and tyrosine kinase signaling pathways (Kretzschmar *et al.*, 1997; de Caestecker *et al.*, 1998). Dissecting these regulatory networks will require finer control of gene expression than is currently obtained by microinjection experiments. Thus, it will be important to isolate tissue-specific promoters and analyze overexpression or loss-of-function phenotypes in transgenic embryos. Further insights into the regulatory pathways governing blood development will come from mutagenesis screens in zebrafish. By a combination of these approaches, it should be possible to determine the genetic pathways underlying early hematopoiesis and thereby decipher the molecular formula for turning mesoderm into blood.

References

Casey, E. S., Tada, M., Fairclough, L., Wylie, C. C., Heasman, J., and Smith, J. C. (1999). *Development* **126**, 4193–4200. *Bix4* is activated directly by VegT and mediates endoderm formation in *Xenopus* development.

Cleaver, O., Tonissen, K. F., Saha, M. S., and Krieg, P. A. (1997). Neovascularization of the *Xenopus* embryo. *Dev. Dyn.* **210**, 66–77.

Clement, J. H., Fettes, P., Knochel, S,. Lef, J., and Knochel, W. (1995). Bone morphogenetic protein 2 in the early development of *Xenopus laevis. Mech. Dev.* **52**, 357–370.

Dale, L., Howes, G., Price, B. M., and Smith, J. C. (1992). Bone morphogenetic protein 4: A ventralizing factor in early *Xenopus* development. *Development* **115**, 573–585.
Dale, L., and Slack, J. M. W. (1987a). Fate map for the 32-cell stage of *Xenopus laevis*. *Development* **99**.
Dale, L., and Slack, J. M. W. (1987b). Regional specification within the mesoderm of early embryos of *Xenopus laevis*. *Development* **100**, 279–295.
D'Andrea, A. D. (1994). Hematopoietic growth factors and the regulation of differentiative decisions. *Curr. Biol.* **6**, 804–808.
de Caestecker, M. P., Parks, W. T., Frank, C. J., Castagnino, P., Bottaro, D. P., Roberts, A. B., and Lechleider, R. J. (1998). Smad2 transduces common signals from receptor serine-threonine and tyrosine kinases. *Genes Dev.* **12**, 1587–1592.
Dick, A., Hild, M., Bauer, H., Imai, Y., Maifeld, H., Schier, A. F., Talbot, W. S., Bouwmeester, T., Hammerschmidt, M. (2000). Essential role of Bmp7 (snailhouse) and its prodomain in dorsoventral patterning of the zebrafish embryo. *Development* **127**, 343–354.
Dosch, R., Gawantka, V., Delius, H., Blumenstock, C., and Niehrs, C. (1997). Bmp-4 acts as a morphogen in dorsoventral mesoderm patterning in *Xenopus*. *Development* **124**, 2325–2334.
Driever, W. (1995). Axis formation in zebrafish. *Curr. Opin. Genet. Dev.* **5**, 610–618.
Ecochard, V., Cayrol, C., Rey, S., Foulquier, F., Caillol, D., Lemaire, P., and Duprat, A. M. (1998). A novel *Xenopus Mix*-like gene *milk* involved in the control of the endomesodermal fates. *Development* **125**, 2577–2585.
Fainsod, A., Steinbeisser, H., and De Robertis, E. M. (1994). On the function of BMP-4 in patterning the marginal zone of the *Xenopus* embryo. *EMBO J.* **13**, 5015–5025.
Fisher, S,. Amacher, S. L., and Halpern, M. E. (1997). Loss of *cerebum* function ventralizes the zebrafish embryo. *Development* **124**, 1301–1311.
Friedle, H., Rastegar, S., Paul, H., Kaufmann, E., and Knöchel, W. (1998). Xvent-1 mediates BMP-4 induced suppression of the dorsal-lip-specific early response gene *XFD-1'* in *Xenopus* embryos. *EMBO J.* **17**, 2298–2307.
Gawantka, V., Delius, H., Hirschfeld, K., Blumenstock, C., and Niehrs, C. (1995). Antagonizing the Spemann organizer: Role of the homeobox gene Xvent-1. *EMBO J.* **14**, 6268–6279.
Gering, M., Rodaway, A. R. F., Göttgens, B., Patient, R. K., and Green, A. R. (1998). The *SCL* gene specifies haemangioblast development from early mesoderm. *EMBO J.* **17**, 4029–4045.
Graff, J. M. (1997). Embryonic patterning: To BMP or not to BMP, that is the question. *Cell* **89**, 171–174.
Graff, J. M., Thies, R. S., Song, J. J., Celeste, A. J., and Melton, D. A. (1994). Studies with a *Xenopus* BMP receptor suggest that ventral mesoderm-inducing signals override dorsal signals in vivo. *Cell* **79**, 169–179.
Green, T. (1996). Master regulator unmasked. *Nature* **383**, 575–577.
Harland, R., and Gerhart, J. (1997). Formation and function of Spemann's organizer. *Annu. Rev. Cell Dev. Biol.* **13**, 611–617.
Heasman, J. (1997). Patterning the *Xenopus* blastula. *Development* **124**, 4179–4191.
Hemmati-Brivanlou, A., and Thomsen, G. H. (1995). Ventral mesodermal patterning in *Xenopus* embryos: Expression patterns and activities of BMP-2 and BMP-4. *Dev. Genet.* **17**, 78–89.
Henry, G. L., and Melton, D. A. (1998). *Mixer*, a homeobox gene required for endoderm development. *Science* **281**, 91–96.
Hild, M., Dick, A., Rauch, G.-J., Meier, A., Bouwmeester, T., Hafter, P., and Hammerschmidt, M. (1999). The *smad5* mutation *somitabun* blocks Bmp2b signaling during early dorsoventral patterning of the zebrafish embryo. *Development* **126**, 2149–2159.
Ho, R. K., and Kane, D. A. (1990). Cell-autonomous action of zebrafish *spt-1* mutation in specific mesodermal precursors. *Nature* **348**, 728–730.
Hoppler, S., and Moon, R. T. (1998). BMP-2/-4 and Wnt8 cooperatively pattern the *Xenopus* mesoderm. *Mech. Dev.* **71**, 119–129.

2. Development of Hematopoietic Stem Cells

Iemura, S.-I., Yamamoto, T. S., Takagi, C., Uchiyama, H., Natsume, T., Shimasaki, S., Sugino, H., and Ueno, N. (1998). Direct binding of follistatin to a complex of bone morphogenetic protein and its receptor inhibits ventral and epidermal cell fates in early *Xenopus* embryo. *Proc. Natl. Acad. Sci. USA* **95,** 9337–9342.

Jones, C. M., Lyons, K. M., Lapan, P. M., Wright, C. V., and Hogan, B. L. (1992). DVR-4 (bone morphogenetic protein-4) as a posterior-ventralizing factor in *Xenopus* mesoderm induction. *Development* **115,** 639–647.

Jones, C. M., and Smith, J. C. (1998). Establishment of a BMP-4 morphogen gradient by long-range inhibition. *Dev. Biol.* **194,** 12–17.

Kelley, C., Yee, K., Harland, R., and Zon, L. I. (1994). Ventral expression of GATA-1 and GATA-2 in the *Xenopus* embryo defines induction of hematopoietic mesoderm. *Dev. Biol.* **165,** 193–205.

Kimmel, C. B., Kane, D. A., Walker, C., Warga, R. M., and Rothman, M. B. (1989). A mutation that changes cell movement and cell fate in the zebrafish embryo. *Nature* **337,** 358–362.

Kishimoto, Y., Lee, K. H., Zon, L., Hammerschmidt, M., and Schulte-Merker, S. (1997). The molecular nature of zebrafish *swirl*: BMP2 function is essential during early dorsoventral patterning. *Development* **124,** 4457–4466.

Kretzschmar, M., Doody, J., and Massague, J. (1997). Opposing BMP and EGF signalling pathways converge on the TGF-β family mediator Smad1. *Nature* **389,** 618–622.

Kumano, G., Belluzzi, L., and Smith, W. C. (1999). Spatial and Temporal properties of ventral blood island induction in *Xenopus laevis*. *Development* **126,** 5327–5337.

Ladher, R., Mohun, T. J., Smith, J. C., and Snape, A. M. (1996). *Xom*: a *Xenopus* homeobox gene that mediates the early effects of BMP-4. *Development* **122,** 2385–2394.

Lane, M. C., and Smith, W. C. (1999). The origins of primitive blood in *Xenopus*: Implications for axial patterning. *Development* **126,** 423–434.

Latinkic, B. V., and Smith, J. C. (1999). *Goosecoid* and *Mix.1* repress *Brachyury* expression and are required for head formation in *Xenopus*. *Development* **126,** 1769–1779.

Lemaire, P., Darras, S., Caillol, D., and Kodjabachian, L. (1998). A role for the vegetally expressed *Xenopus* gene *Mix.1* in endoderm formation and in the restriction of mesoderm to the marginal zone. *Development* **125,** 2371–2380.

Liao, E. C., Paw, B. H., Oates, A. C., Pratt, S. J., Postlethwait, J. H., and Zon, L. I. (1998). SCL/Tal-1 transcription factor acts downstream of *cloche* to specify hematopoietic and vascular progenitors in zebrafish. *Genes Dev.* **12,** 621–626.

Maeno, M., Mead, P. E., Kelley, C., Xu, R. H., Kung, H. F., Suzuki, A., Ueno, N., and Zon, L. I. (1996). The role of BMP-4 and GATA-2 in the induction and differentiation of hematopoietic mesoderm in *Xenopus laevis*. *Blood* **88,** 1965–1972.

Maeno, M., Ong, R. C., and Kung, H.-F. (1992). Positive and negative regulation of the differentiation of ventral mesoderm for erythrocytes in *Xenopus*. *Dev. Growth Differ.* **34,** 567–577.

Maeno, M., Ong, R. C., Xue, Y., Nishimatsu, S., Ueno, N., and Kung, H.-F. (1994). Regulation of primary erythropoiesis in the ventral mesoderm of *Xenopus* gastrula embryo: Evidence for the expression of a stimulatory factor(s) in animal pole tissue. *Dev. Biol.* **161,** 522–529.

Martinez-Barbera, J. P., Toresson, H., Da Rocha, S., and Krauss, S. (1997). Cloning and expression of three members of the zebrafish Bmp family: Bmp2a, Bmp2b and Bmp4. *Gene* **198,** 53–59.

Mayani, H., Guilbert, L. J., and Janowska-Wieczorek, A. (1992). Biology of the hemopoietic microenvironment. *Eur. J. Haematol.* **49,** 225–233.

Mead, P., Kelley, C. M., Hahn, P. S., Piedad, O., and Zon, L. I. (1998a). SCL specifies hematopoietic mesoderm in *Xenopus* embryos. *Development* **125,** 2611–2620.

Mead, P. E., Brivanlou, I. H., Kelley, C. M., and Zon, L. I. (1996). BMP-4-responsive regulation of dorsal-ventral patterning by the homeobox protein Mix.1. *Nature* **382,** 357–360.

Mead, P. E., Zhou, Y., Lustig, K. D., Huber, T. L., Kirschner, M. W., and Zon, L. I. (1998b). Cloning of Mix-related homeodomain proteins using fast retrieval of gel shift activities, (FROGS), a technique for the isolation of DNA-binding proteins. *Proc. Natl. Acad. Sci. USA* **95,** 11251–11256.

Melby, A. E., Clements, W. K., and Kimelman, D. (1999). Regulation of dorsal gene expression in *Xenopus* by the ventralizing homeodomain gene *Vox*. *Dev. Biol.* **211,** 293–305.

Moody, S. A. (1987a). Fates of the blastomeres of the 16-cell stage *Xenopus* embryo. *Dev. Biol.* **119,** 560–578.

Moody, S. A. (1987b). Fates of the blastomeres of the 32-cell-stage *Xenopus* embryo. *Dev. Biol.* **122,** 300–319.

Mullins, M. C., Hammerschmidt, M., Kane, D. A., Odenthal, J., Brand, M., van Eeden, F. J., Furutani-Seiki, M., Granato, M., Haffter, P., Heisenberg, C. P., Jiang, Y. J., Kelsh, R. N., and Nusslein-Volhard, C. (1996). Genes establishing dorsoventral pattern formation in the zebrafish embryo: The ventral specifying genes. *Development* **123,** 81–93.

Neave, B., Holder, N., and Patient, R. (1997). A graded response to BMP-4 spatially coordinates patterning of the mesoderm and ectoderm in the zebrafish. *Mech. Dev.* **62,** 183–195.

Nguyen, V. H., Schmid, B., Trout, J., Connors, S. A., Ekker, M., and Mullins, M. C. (1998). Ventral and lateral regions of the zebrafish gastrula, including the neural crest progenitors, are established by a *bmp2b/swirl* pathway of genes. *Dev. Biol.* **199,** 93–110.

North, T., Gu, T. L., Stacy, T., Wang, Q., Howard, L., Binder, M., Marin-Padilla, M., and Speck, N. A. (1999). Cbfa2 is required for the formation of intra-aotic hematopoietic clusters. *Development* **126,** 2563–2575.

Onichtchouk, D., Gawantka, V., Dosch, R., Delius, H., Hirschfeld, K., Blumenstock, C., and Niehrs, C. (1996). The *Xvent-2* homeobox gene is part of the BMP-4 signalling pathway controlling dorsoventral patterning of *Xenopus* mesoderm. *Development* **122,** 3045–3053.

Papalopulu, N., and Kintner, C. (1996). A *Xenopus* gene, *Xbr-1*, defines a novel class of homeobox genes and is expressed in the dorsal ciliary margin of the eye. *Dev. Biol.* **174,** 104–114.

Parker, L., and Stainier, D. Y. R. (1999). Cell-autonomous and non-autonomous requirements for the zebrafish gene *cloche* in hematopoiesis. *Development* **126,** 2643–2651.

Piccolo, S., Agius, E., Leyns, L., Bhattacharyya, S., Grunz, H., Bouwmeester, T., and De Robertis, E. M. (1999). The head inducer Cerberus is a multifunctional antagonist of nodal, BMP and Wnt signals. *Nature* **397,** 707–710.

Piccolo, S., Sasai, Y., Lu, B., and De Robertis, E. M. (1996). Dorsoventral patterning in *Xenopus*: Inhibition of ventral signals by direct binding of chordin to BMP-4. *Cell* **86,** 589–598.

Porcher, C., Swat, W., Rockwell, K., Fujiwara, Y., Alt, F. W., and Orkin, S. H. (1996). The T cell leukemia oncoprotein SCL/tal-1 is essential for development of all hematopoietic lineages. *Cell* **86,** 47–57.

Ransom, D. G., Haffter, P., Odenthal, J., Brownlie, A., Vogelsang, E., Kelsh, R. N., Brand, M., van Eeden, F. J., Furutani-Seiki, M., Granato, M., Hammerschmidt, M., Heisenberg, C. P., Jiang, Y. J., Kane, D. A., Mullins, M. C., and Nusslein-Volhard, C. (1996). Characterization of zebrafish mutants with defects in embryonic hematopoiesis. *Development* **123,** 311–319.

Robb, L., and Begley, C. G. (1997). The SCL/TAL1 gene: Roles in normal and malignant haematopoiesis. *Bioessays* **19,** 607–613.

Robb, L., Elwood, N. J., Elefanty, A. G., Kontgen, F., Li, R., Barnett, L. D., and Begley, C. G. (1996). The *scl* gene product is required for the generation of all hematopoietic lineages in the adult mouse. *EMBO J.* **15,** 4123–4129.

Rosa, F. M. (1989). Mix.1, a homeobox mRNA inducible by mesoderm inducers, is expressed mostly in the presumptive endodermal cells of *Xenopus* embryos. *Cell* **57,** 965–974.

Sasai, Y., Lu, B., Steinbeisser, H., and De Robertis, E. M. (1995). Regulation of neural induction by the Chd and Bmp-4 antagonistic patterning signals in *Xenopus*. *Nature* **376,** 333–336.

Schmid, B., Furthauer, M., Connors, S. A., Trout, J., Thisse, B., Thisse, C., and Mullins, M. C. (2000). Equivalent genetic roles for *bmp7/snailhouse* and *bmp2b/swirl* in dorsoventral pattern formation. *Development* **127,** 957–967.

2. Development of Hematopoietic Stem Cells

Schmidt, J. E., von Dassow, G., and Kimelman, D. (1996). Regulation of dorsal-ventral patterning: The ventralizing effects of the novel *Xenopus* homeobox gene *Vox. Development* **122,** 1711–1721.

Schulte-Merker, S., Lee, K. J., McMahon, A. P., and Hammerschmidt, M. (1997). The zebrafish organizer requires *chordino. Nature* **387,** 862–863.

Sive, H. L. (1993). The frog prince-ss: A molecular formula for dorsoventral patterning in *Xenopus. Genes Dev.* **7,** 1–12.

Smith, J. C. (1989). Mesoderm induction and mesoderm-inducing factors in early amphibian development. *Development* **105,** 665–677.

Smith, J. C. (1995). Mesoderm-inducing factors and mesodermal patterning. *Curr. Opin. Cell Biol.* **7,** 856–861.

Solnica-Krezel, L. (1999). Pattern formation in zebrafish–Fruitful liaisons between embryology and genetics. *Curr. Top. Dev. Biol.* **41,** 1–35.

Spemann, H., and Mangold, H. (1924). Uber Induktion von Embryonenanlagen durch Implantation artfremder Organisatoren. *Wilhelm Roux' Arch. EntwMech. Org.* **100,** 599–638.

Stainier, D. Y., Weinstein, B. M., Detrich, H. W., 3rd, Zon, L. I., and Fishman, M. C. (1995). *Cloche*, an early acting zebrafish gene, is required by both the endothelial and hematopoietic lineages. *Development* **121,** 3141–3150.

Steinbeisser, H., Fainsod, A., Niehrs, C,. Sasai, Y,. and De Robertis, E. M. (1995). The role of gsc and BMP-4 in dorsal-ventral patterning of the marginal zone in *Xenopus*: A loss-of-function study using antisense RNA. *EMBO J.* **14,** 5230–5243.

Suzuki, A,. Thies, R. S., Yamaji, N., Song, J. J., Wozney, J. M., Murakami, K., and Ueno, N. (1994). A truncated bone morphogenetic protein receptor affects dorsal-ventral patterning in the early *Xenopus* embryo. *Proc. Natl. Acad. Sci. USA* **91,** 10255–10259.

Tada, M., Casey, E. S., Fairclough, L., and Smith, J. C. (1998). *Bix1*, a direct target of *Xenopus* T-box genes, causes formation of ventral mesoderm and endoderm. *Development* **125,** 3997–4006.

Thompson, M. A., Ransom, D. G., Pratt, S. J., MacLennan, H., Kieran, M. W., Detrich, H. W., Vail, B., Huber, T. L., Paw, B., Brownlie, B., Oates, A. J., Fritz, A., Gates, M. A., Amores, A., Bahary, N., Talbot, W. S., Her, H., Beier, D. R., Postlethwait, J. H., and Zon, L. I. (1998). The *cloche* and *spadetail* genes differentially affect hematopoiesis and vasculogenesis. *Dev. Biol.* **197,** 248–269.

Thomsen, G. H. (1997). Antagonism within and around the organizer: BMP inhibitors in vertebrate body patterning. *Trends Genet.* **13,** 209–211.

Tidman Ault, K., Dirksen, M.-L., and Jamrich, M. (1996). A novel homeobox gene *PV.1* mediates induction of ventral mesoderm in *Xenopus* embryos. *Proc. Natl. Acad. Sci. USA* **93,** 6415–6420.

Tracey, W. D. J., Pepling, M. E., Horb, M. E., Thomsen, G. H., and Gergen, J. P. (1998). A *Xenopus* homologue of *aml-1* reveals unexpected patterning mechanisms leading to the formation of embryonic blood. *Development* **125,** 1371–1380.

Turpen, J. B., Kelley, C. M., Mead, P. E., and Zon, L. I. (1997). Bipotential primitive-definitive hematopoietic progenitors in the vertebrate embryo. *Immunity* **7,** 325–334.

Vize, P. D. (1996). DNA sequences mediating the transcriptional response of the *Mix.2* homeobox gene to mesoderm induction. *Dev. Biol.* **177,** 226–231.

Vodicka, M. A., and Gerhart, J. C. (1995). Blastomere derivation and domains of gene expression in the Spemann organizer of *Xenopus laevis. Development* **121,** 3505–3518.

Wang, S., Krinks, M., Kleinwaks, L., and Moos, M. J. (1997). A novel *Xenopus* homologue of bone morphogenetic protein-7 (BMP-7). *Genes Funct.* **1,** 259–271.

Whitman, M. (1998). Smads and early developmental signaling by the TGFβ superfamily. *Genes Dev.* **11,** 2445–2460.

Wilson, D., Sheng, G., Lecuit, T., Dostatni, N., and Desplan, C. (1993). Cooperative dimerization of paired class homeo domains on DNA. *Genes Dev.* **7,** 2120–2134.

Xu, R.-H., Tidman Ault, K., Kim, J., Park, M.-J., Hwang, Y.-S., Peng, Y., Sredni, D., and Kung, H.-F. (1999). Opposite effects of FGF and BMP-4 on embryonic blood formation: Roles of PV.1 and GATA-2. *Dev. Biol.* **208,** 352–361.

Zhang, J., King, M. L., Houston, D., Payne, C., Wylie, C., and Heasman, J. (1998). The role of maternal VegT in establishing the primary germ layers in *Xenopus* embryos. *Cell* **94,** 515–524.

Zimmerman, L. B., De Jesus-Escobar, J. M., and Harland, R. M. (1996). The Spemann organizer signal noggin binds and inactivates bone morphogenetic protein 4. *Cell* **86,** 599–606.

3
Mechanisms of Plant Embryo Development

Shunong Bai, Lingjing Chen, Mary Alice Yund, and Zinmay Renee Sung
Department of Plant and Microbial Biology
University of California
Berkeley, California 94720

I. Introduction
II. Embryogenesis
    A. Growth and Morphogenesis Are Gradual and Continuous
    B. Embryo and Seedling Development Is a Continuous Process Interrupted by Seed Development
III. Histogenesis: The Generation of Cells and Tissues
    A. Are There Maternal Effects on Plant Embryo Development?
    B. How Are Cell Fates Specified?
    C. Molecular Mechanism of Radial Patterning
    D. Polar Auxin Transport as a Signaling Mechanism for Procambium Differentiation
IV. Organogenesis: The Generation of Organs
    A. Embryonic Polarity and Axial Patterning in Dicot Embryos
    B. Root Meristem Ontogeny and Function
    C. Shoot Meristem Ontogeny and Function
    D. Mechanism of Cotyledon Formation
    E. A Nongenetic Mechanism
V. Embryo-Specific Genes and Embryonic Mutants
VI. Embryonic Induction
    A. Zygotic Embryogenesis
    B. Adventive Embryogenesis from Ovary Tissue
    C. Somatic Embryogenesis from Callus
    D. Changing Embryogenic Competence during the Plant Life Cycle
    E. Acquisition of Embryonic Competence during Reproductive Growth
VII. Summary
    References

I. Introduction

Plant developmental biology investigates how a single cell, the zygote, gives rise to a multicellular organism. The development of flowering plants begins with formation of the organism in the seed, the embryo, which is much simpler in structure than the adult plant. Most embryos are composed of embryonic leaves and a short axis, on each end of which is a primary meristem. Flower formation and other complex morphogenetic processes of plants do not occur during embryogenesis and cannot be studied by an analysis of embryogenesis.

As a corollary of this embryonic simplicity, early plant development is very uniform across species and phyla. Adult plants are as diverse as *Arabidopsis*, a weed several inches high, and *Eucalyptus*, a tree several hundred feet in height, but the embryos of all monocots have one cotyledon whereas all dicots have two. This embryonic similarity suggests that the early, fundamental process of embryogenesis has been conserved during evolution and that similar principles or mechanisms may direct plant embryo development in diverse species. Developmental diversity occurs at later stages in the plant life cycle. Thus the study of plant embryogenesis holds particular promise for revealing principles and mechanisms of plant development. Studies of plant embryos provide a window into such fundamental questions as regulation of cell proliferation, pattern formation, and differentiation. This review analyzes plant embryo development from the point of view of morphogenesis, histogenesis, and organogenesis.

More specifically, we will address the following questions:

1. How does cell differentiation take place and how are cell fates determined?
2. How are tissues organized—radial patterning?
3. How are organs specified—axial patterning?
4. How are shoot and root apical meristems formed?
5. How is the embryogenic fate induced?

II. Embryogenesis

A. Growth and Morphogenesis Are Gradual and Continuous

Embryogenesis begins with fertilization. Growth and morphogenesis are continuous as the embryo changes in size and shape (Fig. 1A). Division of the zygote generates a ball of cells that show preferential growth at the apical end. Growth at the center of the apical end of the dicot embryo is slower than at the periphery, generating a depression in the spherical structure. Later growth is restricted at opposite sides of the periphery, resulting in the formation of two cotyledons. The sequential stages of the developing embryo are often described as shapes: globular, heart, and torpedo. Embryonic forms and organogenesis are closely correlated with the pattern of cell proliferation and cell elongation. The principles and underlying forces, biochemical or physical, regulating embryo metamorphosis are poorly understood (Green, 1992). The causal relationship between morphogenesis and histogenesis is discussed in Kaplan and Cooke (1997).

3. Mechanisms of Plant Embryo Development 63

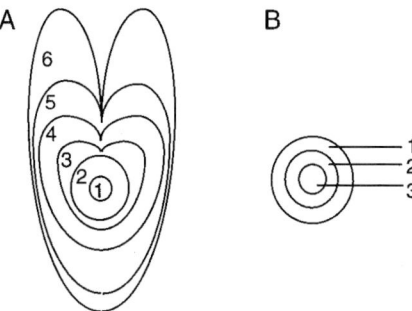

Fig. 1 Schematic representation of continuous growth and development of a dicot embryo proper (A) and of histogenesis (B). (A) 1, zygote; 2, globular stage; 3, early heart stage; 4, heart stage; 5, early torpedo stage; and 6, torpedo stage. (B) Cross section of the embryonic axis showing the radial patterning of the three fundamental tissues in a late globular stage *Arabidopsis* embryo: 1, protoderm; 2, ground tissue; and 3, procambium.

B. Embryo and Seedling Development Is a Continuous Process Interrupted by Seed Development

Producing a seed requires activation of the genes responsible for seed maturation processes that include production of the food stored in the seed, desiccation, dormancy, and eventual seed germination (Bewley and Black, 1994). These seed maturation processes are superimposed on embryogenesis and are coordinated with the progression of embryo development. However, there is great variation among plants in the relationship between seed formation and embryogenesis. Embryo development arrests when a seed is formed and resumes after germination. Most dicotyledon embryos arrest growth and morphogenesis, including apical meristem activities, after the torpedo stage. *Phaseolus* and many monocotyledon embryos develop several true leaves before seed maturation (Fig. 2A). On the other hand, orchid seed form and dessicate before cotyledon differentiation begins (Fig. 2B). Embryo development continues during seed imbibition (Rao, 1967; Poddubnaya-Arnoldi, 1967).

Both seed storage proteins and late embryo abundant (LEA) proteins of unknown function are produced by developing embryos. The espression of these proteins in embryo suggests that they are regulated by an embryonic program and that the temporal and spatial expression of seed maturation proteins are useful markers of embryogenesis. This was, in part, because of the apparent embryo-specific expression of these genes in nonendospermatous seeds. However, in endospermatous seeds, these genes are often expressed in both embryo and endosperm (Fig. 3) (Borkird *et al.*, 1988; Goupil *et al.*, 1992). Close examination of the temporal and spatial expression of such genes in plants and in mutants with altered or aberrant embryogenesis revealed that the seed maturation pro-

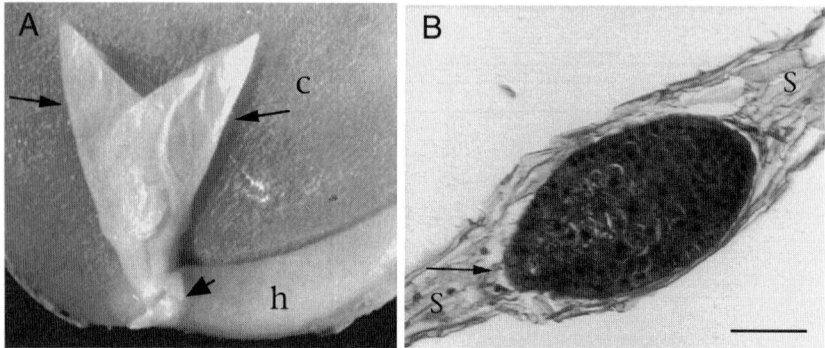

Fig. 2 *Phaseolus* embryos (A) showing precocious development of true leaves and *Bletilla ochracea* embryos (B) showing precocious cessation of organogenesis in mature seeds. (A) c, cotyledon; h, hypocotyl; long arrows point to two true leaves with a well-developed venation pattern; short arrow points to site where the other cotyledon was removed. (B) s, seed coat; arrow points to the micropylous end of the embryo; the section is 10 μm in thickness; nuclei were stained with hematoxylin. Bar: 100 μm.

gram is coordinated with, but separable from, embryogenesis (Cheng *et al.*, 1996; Yadegari *et al.*, 1994). Expression of LEA genes and seed development take place in mutants with defective embryos arrested at early development (Goldberg *et al.*, 1989). Conversely, immature embryos genetically impaired in abscisic acid synthesis and sensitivity necessary for the seed maturation process

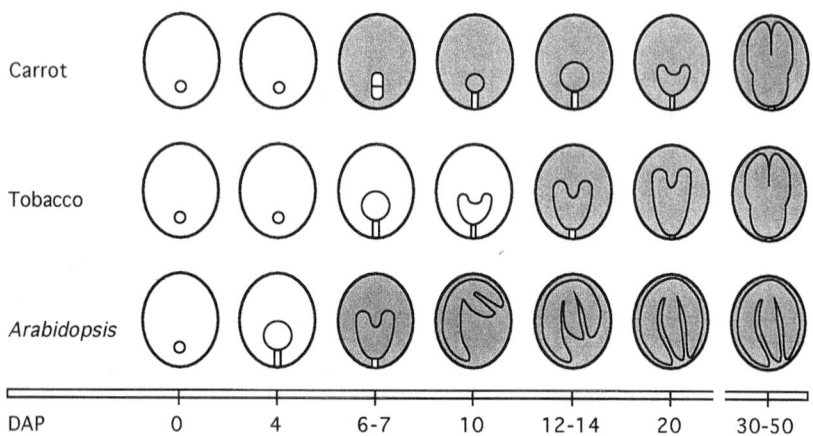

Fig. 3 LEA gene expression in carrot, tobacco, and *Arabidopsis*. Diagrammatic representation of embryo and endosperm development of carrot, tobacco, and *Arabidopsis* with the temporal expression of the LEA gene, DC8::GUS construct, depicted by shading. Seeds are not drawn to scale. (Reprinted from *Plant Molecular Biology*, **31**, 1996, pp. 127–41, Expression of DC8 is associated with, but not dependent on embryogenesis, Cheng *et al.*, Figure 3 with kind permission from Kluwer Academic Publishers.) DAP, days after pollination.

can develop viviparously inside siliques on plants or when young siliques are imbibed on wet filter paper (Koornneef *et al.*, 1989). These observations support the separateness of the embryogenetic and seed maturation processes.

III. Histogenesis: The Generation of Cells and Tissues

Embryo histogenesis usually begins with unequal cleavage of the zygote generating a smaller, upper cell rich in cytoplasm and a larger, lower cell that contains a large vacuole and is poorer in cytoplasm. The upper cell continues to divide rapidly to form the globular embryo proper. The larger lower cell makes a few transverse divisions to form the suspensor structure that connects the embryo to the maternal tissue. The uppermost cell generated from division of the suspensor-forming cell from the first cleavage is called the hypophysis. This cell divides a few times to form a small wedge of cells that become part of the embryonic root (Johansen, 1950, Fig. 5).

At the beginning of embryogenesis, during the cleavage period, embryonic cell number increases as the average cell size decreases (Pollock and Jensen, 1964). Cell division is rapid, oscillating between mitosis and DNA synthesis bypassing G1 and G2. Continuing growth of the embryo results in the formation of three fundamental tissues: (1) protoderm, the surface layer of cells formed early in the globular stage that later differentiates into the epidermis, (2) procambium, narrow cells in the embryo center visible by the late heart stage that later differentiate into the vascular tissue, and (3) ground tissue, the tissue between protoderm and procambium that later differentiates into the cortical tissue (Fig. 1B).

A. Are There Maternal Effects on Plant Embryo Development?

Plant embryos can differentiate from callus tissue derived from leaf, root, and other tissues, a phenomenon commonly referred to as somatic embryogenesis (see later), or from cells other than the egg, e.g., hypocotyl or nucellus cells of *Ranuculus* (Konar and Nataraja, 1965). Thus, embryogenesis in plants does not seem to be dependent on maternal information. However, the phenotypes of two *Arabidopsis* mutants, *short integument 1 (sin1)* and *medea*, could be caused by the maternal influence on embryo development.

The *sin1* mutant displays a variety of phenotypes, including aberrant ovule development, which could be caused by a defect in a diffusible product from the maternal tissue necessary for embryonic pattern formation (Ray *et al.*, 1996). Alternatively, the various abnormal embryonic forms could result from structural defects in the ovule and embryo sac formed in homozygous *sin1* flowers.

The phenotype of the *medea* mutant suggests maternal control of embryonic cell proliferation (Grossniklaus *et al.*, 1998). A maternally inherited mutant

MEDEA allele causes excessive cell proliferation and an increased number of cell files in the embryo and an enlarged embryo with extra cells. The mutation does not affect the morphogenetic progression of the embryo, but the excessive and lengthened proliferative activity in the embryo delays cellularization in the endosperm and affects the seed maturation process. *medea* mutant embryos degenerate during desiccation, but can continue to develop through embryo rescue. A maternal affect on embryo size is consistent with the observation that somatic embryos are often abnormal in size and shape. However, such abnormalities usually do not prevent the somatic embryo from forming adult plants.

B. How Are Cell Fates Specified?

Does an asymmetrical zygotic division generate cells of different fates? The zygote of the typical dicot *Capsella* is an elongated cell with the nucleus at the apical end and a large vacuole at the basal end (Fig. 4B). Given this cytology, the first zygotic division is asymmetric, creating a smaller, cytoplasm-rich apical cell that gives rise to almost all of the embryo and a larger, vacuolated basal

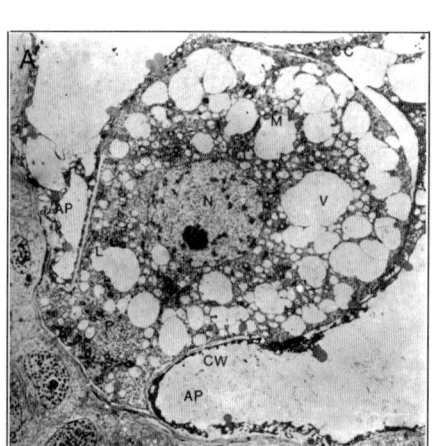
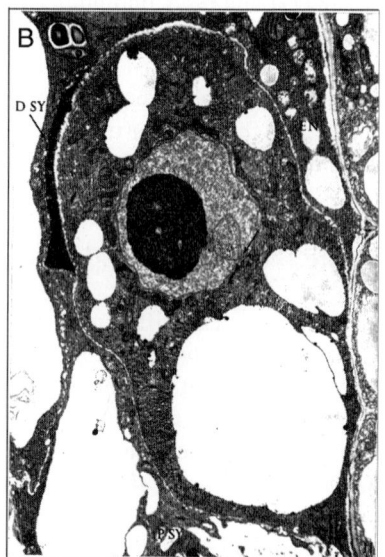

Fig. 4 Transmission electron micrographs of a wheat egg cell (A) and a *Capsella* zygote (B). (A) Wheat egg cell showing centrally located nucleus and small vacuoles distributed evenly throughout the cytoplasm. AP, apical pocket; CC, central cell; CW, cell wall; M, mitochondria; N, nucleus (from You and Jensen, 1985). (B) The zygote of *Capsella* showing a nucleus at the chalazal end and a large vacuole at the micropylous end. A small segment of the degenerating synergid (DSY) can be seen, as well as some of the endosperm (EN). From Schulz and Jensen (1968) with permission.

suspensor cell that forms the link with the maternal tissue and contributes only a few cells to the embryo proper. These two cells are clearly cytologically different and give rise to functionally diverse tissues. Cells of very different fates appear to be created at the first division. The daughter cells could have received different morphological determinants from the zygotic cell, causing differential gene expression. An asymmetric cell division could be critical to cell differentiation (Wardlaw, 1953).

Alternatively, the distinct cytology of the two cells created at the asymmetric division may not have committed the cells to different fates. As the embryo grows and develops, these cells and their descendants interact with each other. Each cell could interpret its position in the embryo with respect to other cells, resulting in the gradual emergence of the differentiation pattern.

Several pieces of evidence support the notion that this asymmetrical division does not play a definitive role in cell fate diversification.

First, not all plant species have eggs with visible cytoplasmic asymmetry and unequal zygotic division. Examination of a wheat embryo does not suggest an embryo/suspensor distinction at the first division. Wheat egg cells appear to be apolar (Fig. 4A), do not contain a large vacuole (You and Jensen, 1985), and undergo equal division, producing two cells that appear to contain the same amount of cytoplasm. Nevertheless, subsequent growth and division lead to the formation of an embryo and a suspensor. The zygotic division of *Phaseolus* is asymmetric, but both daughter cells divide to form a filamentous proembryo with little structural differentiation. Cells derived from terminal and basal cells appear to have the same general cytological features (Yeung and Clutter, 1978). These observations are consistent with suspensor/embryo differentiation taking place during subsequent embryonic growth.

Second, *Arabidopsis* mutants *sus2* (Schwartz et al., 1994) and raspberry (Yadegari et al., 1994) produce embryos from the suspensor. Zygotes of these mutants undergo asymmetric division, but when embryonic growth is arrested by mutation, the suspensor cells divide to produce a second proembryo. Apparently the suspensor cells are still embryogenic after asymmetrical cell division, but the embryogenic potential is suppressed by the growth of the embryo from the apical cell. Other mutants, *twn1* and *twn2*, produce normal embryos from the apical cell and additional embryos from the suspensor, resulting in polyembryony and twin seedlings at germination (Vernon and Meinke, 1994; Zhang and Sommerville, 1997). The *twn2* mutation, a defect in valyl-tRNA synthetase, provides no insight into the molecular mechanism of cell fate determination, but does demonstrate that the suspensor cell can produce an embryo. The asymmetrical first division did not cause or correspond to an immediate loss of embryogenic potential in the suspensor or commit the cells to different cell fates.

The phenotypes of embryonic mutations such as *sus2* reveal that the apparent fate of apical and suspensor cells, while predictable, is presumptive and not irrevocably set at the first division. Furthermore, embryogenic potential extends beyond those cells that normally form the embryo proper. The apparent inhibitory effect of the developing embryo on the embryogenic potential of the suspensor cell indicates that cell–cell signaling plays a role in cell fate diversification. We do not know why the embryogenic potential becomes inhibited in the suspensor cell but not in the apical cell. It may have to do with the position of the two cells relative to each other and to the surrounding tissue.

If cell fates are not determined at the first zygotic division, how are cell and tissue fates specified during embryogenesis? There are two general types of embryonic histogenesis: (1) an embryo with few cells and a regular, predictable division pattern in which cell lineages can be followed, e.g., *Arabidopsis* and *Capsella;* and (2) an embryo with a large number of cells and a random, irregular, and unpredictable pattern of cell division, e.g., cotton, carrot, and phlox. The cell lineage pattern of these embryos cannot be followed.

Observations resulting from the ability to trace cell lineages in small embryos led to the initial cell concept, the teleological supposition that morphogenetic determinants are segregated through cell division and that cell fates are determined in the progenitor or initial cells, which then propagate determinants to their descendants (Cutter, 1969; Dolan *et al.*, 1993). Early studies emphasize the role of cell division pattern in plant classification (Johansen, 1950) and cell fate specification (Dolan *et al.*, 1993). However, it is not clear how the cell lineage hypothesis can apply to large embryos with unpredictable division patterns. Furthermore, clonal analysis has always shown that cell position, not cell lineage, controls cell fate (Poethig *et al.*, 1986; Poethig, 1987). Data also indicate that initial cells can be readily replaced (Kaplan and Cooke, 1997; van den Berg *et al.*, 1997). A similar lack of correlation between segmentation pattern and organ formation is observed in Pteridophytes (Wardlaw, 1953). An alternative explanation for the regular cell division pattern in small embryos is biophysical restraints. Cells divide to conform to organ shape (Kaplan, 1984; Kaplan and Hagemann, 1992). The observed pattern is the most energy-efficient, stable form of division. In small embryos that contain few cells there are limited ways cells may divide without upsetting the overall growth dynamics.

C. Molecular Mechanism of Radial Patterning

The protoderm, ground tissue, and procambium are the progenitor tissues of the adult epidermis, cortex, and vascular bundles, respectively. Embryonic tissues contain immature, not fully differentiated, cells. For example, procambial cells are not lignified as mature xylem cells are. Because these tissues are organized into a radial pattern in the plant (Fig. 1B), the term radial patterning is used to describe the molecular events that cause the commitment of cells to differentiate into different tissues. Commitment probably occurs prior to the histological differentiation of the three tissues. Recent data have begun to shed light on the radial patterning process.

3. Mechanisms of Plant Embryo Development

The protoderm of *Capsella* becomes distinct by the 16 cell stage of embryogenesis. Cells in the protoderm, the only cells in direct contact with the external environment, divide anticlinally. Mutants with a fiddlehead phenotype define nine complementation groups that reveal a genetic program important in maintaining the epidermal layer or plant surface. In the absence of one of the *FIDDLEHEAD* gene functions, epidermal cells fuse as they come in contact with each other in a process resembling carpal fusion (Lolle *et al.*, 1998).

The *Arabidopsis* gene *ATML1*, encoding a novel homeodomain protein, is expressed in all apical cells of a proembryo but is restricted to the protoderm by

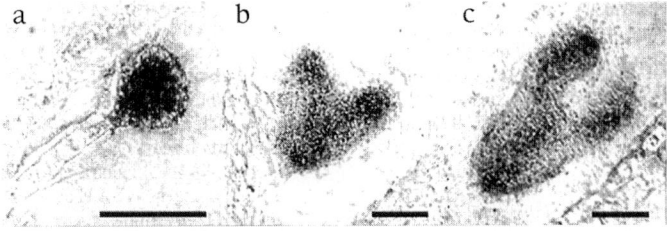

Fig. 5 Expression of *ATML1* (A) and *MP* (B) genes during embryogenesis of *Arabidopsis*. (A) *ATML1* mRNA accumulation during embryo development is shown as darkened areas on the diagrams. Morphological features and stages are as follows: a, apical cell; b, basal cell; c, cotyledon; cc, central cell; col, columella; ep, embryo proper, FE, fertilized egg; GE, globular stage embryo; HE, heart stage embryo; hs, hypophysis; icm, inner cell mass; ME, mature embryo; pd, protoderm; rm, root meristem; s, suspensor; sm, shoot meristem; TE, torpedo stage embryo. (B) Expression pattern of *MP* mRNA in wild-type (Col-0) embryos. *In situ* hybridization with *MP* antisense probe. (a) Early globular stage, (b) early heart stage, and (c) early torpedo stage embryo. Bars are 50 μm. Part A from Lu *et al.* (1996) by permission of the American Society of Plant Physiologists. Part B from Hardtke and Berleth (1998) by permission of Oxford University Press.

the 16 cell stage (Fig. 5A) and its mRNA is absent in the epidermal layer of the mature embryo. This differential expression pattern is maintained in the adult plant where *ATML1* is expressed in the protoderm of the shoot apical meristem but not in the epidermis of the plant. This *ATML1* RNA expression pattern demonstrates molecular differentiation in early embryogenesis and provides a molecular marker for the protoderm. Another protoderm-specific gene, the lipid transfer protein, is expressed in globular stage embryos, the protoderm of the embryonic shoot, the shoot apices of adult carrots (Sterk *et al.*, 1991), and in all cultured carrot cell clusters that develop into an embryo (Toonen *et al.*, 1997).

An interesting ontogenic question has been whether the protoderm gives rise to the ground tissue or vice versa. Examining the surface of the *Citrus* zygote by electron microscopy, Bruck and Walker (1985) reported finding a cuticle layer characteristic of epidermal cells, suggesting that the zygote itself is an epidermal cell and that the 8 cell stage embryo consists of only epidermal cells. Periclinal division of these 8 cells results in a 16 cell stage embryo with 8 surface and 8 internal cells. The latter differentiate into the ground tissue. These observations would suggest that the protoderm is the progenitor of the ground tissue.

The expression pattern of *ATML1* during embryogenesis is consistent with the notion that the protoderm gives rise to the ground tissue. *ATML1* is first expressed in the apical cell following zygotic division and in all embryonic cells until the 16 cell stage when it becomes restricted to the protoderm layer. Ground tissue does not express *ATML1*, perhaps because its cells are no longer in contact with the environment. The RNA of another homeodomain-containing gene, *KNAT1*, has been found to be expressed as early as the late globular stage of the *Arabidopsis* embryo, but only in the ground tissue (Chuck, personal communication). The expression of *ATML1* at an earlier time than *KNAT1* is consistent with protoderm differentiation before ground tissue.

Histologically, the ground tissue appears to be the progenitor of the procambium in the embryo. The earliest sign of procambium differentiation is a change in cell shape and then the procambium becomes distinct from the ground tissue as parallel files of elongated and narrow cell in the center of heart-stage dicot embryos. Following germination, procambial cells differentiate into the xylem and phloem elements that constitute the vascular tissue in the plant.

Procambium-specific genes such as *AthB-8* (Baima *et al.*, 1995) and *MONOPTEROS* (*MP*) (Hardtke and Berleth, 1998) have been identified. The *MP* gene is expressed in all subepidermal cells of young *Arabidopsis* globular embryos. Later expression becomes localized to a more central domain along the midlines of the embryonic axis and the cotyledons (Fig. 5B) and is ultimately confined to provascular tissues of mature embryos and adult plants (Hardtke and Berleth, 1998). *ATML1* and *MP* expression before tissue differentiation is consistent with a role in patterning the tissue differentiation.

D. Polar Auxin Transport as a Signaling Mechanism for Procambium Differentiation

Two pieces of experimental evidence suggest that polar auxin transport provides the signal to initiate procambium differentiation.

First, cells in culture proliferate randomly and form callus tissue when growth hormones in the surroundings flow through the tissue from multiple directions. Normally there is no vascular tissue in the callus, but if a bud is grafted onto cultured callus, vascular tissue differentiates below the bud and extends toward the medium. Further experimentation has revealed that an auxin gradient initiated by auxin moving from the bud into the callus is the stimulus for procambium differentiation (Camus, 1949; Sachs, 1991). Even in adult plants, mature cortical cells can be induced to differentiate into vascular cells in response to a polarized flow of auxin (Sachs, 1991), but the mechanism of redifferentiation is not clear.

Second, in the *Arabidopsis mp* mutant (Przemeck *et al.*, 1996), a row of wide cells stacked on top of one another replaces the files of elongated cells that form during late globular stage embryogenesis. The venation pattern of adult *mp* plants is simple, often with poor cell alignment and differentiation. The *MP* gene encodes a protein with homology to *ARF1* and a DNA-binding domain for the *cis* elements of auxin-inducible genes (Hardtke and Berleth, 1998; Ulmasov *et al.*, 1997). The *mp* mutant has reduced polar auxin transport, suggesting that polar auxin transport could be the primary means by which procambium differentiation is initiated. Polar auxin transportation has been shown to be required for provascular differentiation (Mattsson *et al.*, 1999). These observations suggest that polarized auxin flow through the ground tissue is instrumental in inducing procambium differentiation.

IV. Organogenesis: The Generation of Organs

Key organogenic events in the embryo are specification and differentiation of the embryonic axis and the cotyledons. The organism in the seed appears to have only two parts: cotyledons and the embryonic axis. Germination of the seedling reveals that three organs have been specified during embryogenesis: cotyledons or embryonic leaves, the hypocotyl or embryonic shoot, and the embryonic root. Cells in the embryonic axis are differentially committed to the hypocotyl and root fate along the hypocotyl/root junction (Cheng *et al.*, 1995). Upon germination these fates are expressed, leading to the differentiation of root and hypocotyl. Cells above the hypocotyl/root junction elongate without division and form the green shoot (Cheng *et al.*, 1995). Cells immediately below the junction are not green and form the root of the seedling.

A. Embryonic Polarity and Axial Patterning in Dicot Embryos

Axial patterning is the process by which organ fates are specified along the longitudinal axis of the embryo. Morphologically, polarity can be seen in the early heart stage embryo that grows along an axis and has different ends, with the shoot end being wider than the root end. Histologically, polarity is first seen as elongated cells of the future procambium appear in the globular stage embryo. Because the egg cell of many species appears to be a polarized structure and because the first zygotic division is unequal, it was often assumed that the embryonic polarity that specifies the shoot–root orientation of the embryo, eventually leading to organ differentiation, is established in the egg cell or shortly afterward (Wardlaw, 1953). However, as discussed earlier, embryos can form from vegetative cells or nucellus cells or suspensor cells and develop into polarized organisms. Thus, it appears that without preexisting polarity or unequal cell division, cells committed to embryogenesis can develop axial polarity.

The flow of auxin from the apical to the basal end of the plant, first detected in *Phaseolus* embryos by Fry and Wangerman (1976), has long been recognized as an important physiological and developmental mechanism. This polarized auxin transport may be the basic axial patterning mechanism that establishes the histologically observed embryonic polarity that becomes the adult shoot/root axis. That *mp* mutants impaired in auxin transport lack an embryonic axis also suggests the regulation of axial patterning by polar auxin transport.

B. Root Meristem Ontogeny and Function

Shoot and root apical meristems are identified cytologically as undifferentiated, cytoplasmically rich cells located at the shoot and root apices, respectively. Their primary function, as indicated by the Latin term "meris," a division, is to generate cells. Throughout the life of the plant these new cells differentiate to generate organs. The meristems themselves are maintained as apical cells that remain undifferentiated and competent to divide.

In small embryos with a predictable pattern of cell division, such as *Arabidopsis*, the lineage of cells at the root apex can be traced back to the hypophyseal cell formed a few cell divisions after fertilization. This remarkable observation has been interpreted to suggest that the root apical cell fate was determined in a meaningful, mechanistic way as soon as the hypophyseal cell was formed. The hypophyseal cell was thought to give rise to the "root initial" cell(s) that can renew itself while generating more root apical cells. Thus, the root initial cell was thought to be generated and maintained within the root meristem where it determined the "pattern" of root development (Dolan *et al.*, 1993).

This initial cell concept implies that root initial cells and the root meristem are necessary for root development. If this is true, removal of these cells should render the plant rootless. When the lower portions of carrot somatic embryos were surgically removed, cells at the cut end of the shoot divided and regenerated a root (Schiavone and Racusen, 1991). This shows that the root meristem and the initial cell within it are not necessary for root formation. Furthermore, the cut end regenerated the most proximal tissue, the maturation region, first and replaced more distal tissue later. The root meristem was the last tissue regenerated. Because the root meristem was formed after the root, the meristem cannot be necessary for establishing the pattern of root development. Moreover, that the shoot half of the embryo replaced the missing root suggests that signals from the apical end of the embryo determine the type of organs formed at the basal end (Schiavone and Racusen, 1991).

C. Shoot Meristem Ontogeny and Function

Numerous reviews discuss the definition and function of shoot apical meristem (SAM), its cytological features in relation to leaf initiation and cell division functions, and its role in regulating cell fate (Clark, 1997; Kerstetter and Hake, 1997; Medford, 1992; Miksche and Brown, 1965). Here we will consider the ontogeny and function of the shoot apical meristem in the embryo and review gene expression in the embryonic meristem.

Expression of the homeodomain genes *KNOTTED 1 (KN-1)* of maize (Smith *et al.*, 1995) and *SHOOT MERISTEMLESS (STM)* of *Arabidopsis* (Long *et al.*, 1996) in the apical end of young embryos provides molecular evidence that the shoot apical meristem is specified during the globular stage of early embryogenesis. Possible functions of these genes include (1) promoting cell proliferation, because ectopic expression of *KN-1* in tobacco and *Arabidopsis* causes cell division and formation of new shoots from leaves (Chuck *et al.*, 1996; Sinha *et al.*, 1993); (2) suppressing cell division, because the embryonic shoot meristem region is much less active in cell division than the surrounding cells that produce the cotyledons; (3) maintaining the undifferentiated state, because *stm* mutants are antagonistic to the CLAVATA (CLV) genes that promote differentiation of the meristem (Clark *et al.*, 1996, 1997); and (4) regulating the size of the meristem, together with *CLV*, because although *stm* mutants lack shoot apical meristem and *clv* mutants have an enlarged meristem, the double mutant can appear quite normal.

The "stem cell" or "initial cell" property proposed for meristems (Steeves and Sussex, 1988) would imply that meristematic activity maintains itself or that *STM* gene activity is propagated via cell lineage. The inability to find evidence supporting the "initial cell" theory or cell lineage hypothesis opens the possibility that the cell proliferative property may be regulated by positional information or extracellular factors coming from other parts of the plant. Developmental states (Miksche and Brown, 1965) and environmental factors such as the season

(Kemp, 1943) can regulate the size of the SAM. In tobacco, the developmental state of the SAM is regulated by signals from the root or other parts of the shoot (McDaniel, 1978). This evidence, although circumstantial, nevertheless suggests that *STM* genes may be activated and maintained by extracellular signals generated within the context of the entire plant body.

Indeed, several genes related to SAM development are expressed in provascular or vascular tissues, e.g., *KN-1* (Jackson et al., 1994) and *CLV1* (Jeong and Clark, 1997). The *ZWILLE (ZLL)* gene is expressed in the procambial cells below cells expressing *STM* at globular stage embryogenesis (Moussian et al., 1998). While *ZLL* continues to be expressed in the procambium, after torpedo stage embryogenesis, it is also expressed in the embryonic apex. Both *ZLL* and *MP* RNA are found in the provascular tissues of adult plants. Because *ZLL* mutations cause the shoot apical cells to differentiate, *zll* mutants seldom form leaves after germination. This phenotype led to the proposal that *ZLL* regulates the expression of the *STM* gene in late embryogenesis. The expression of *ZLL* in provascular cells suggests that it may affect *STM* activity by mediating the transport of as yet unknown molecules or plant hormones that regulate shoot meristem activity.

Several other genes expressed during early embryogenesis appear to be related to SAM activity and axial patterning. Examples include *UFO, LEAFY (LFY)*, and *KNAT-1* (Evans and Barton, 1997; Bai et al., unpublished observation; Smith et al., 1995). We observed *LFY* expression as early as globular stage embryos. After the torpedo stage, expression is restricted to the shoot apical region (Fig. 6), consistent with *LFY* expression observed in the SAM and young leaf primordia of adult plants (Blazquez et al., 1997). *LFY* expression is also observed in developing, but not fully differentiated cotyledons and leaves (Fig. 6). The role of these genes during SAM development remains elusive.

Two genes in addition to *STM* produce a similar shoot meristemless phenotype: *WHSCHEL (WUS)* (Laux et al., 1996; Mayer et al., 1998) and *NO APICAL MERISTEM (NAM)* (Souer et al., 1996). *wus-1 stm-2* double mutants display defects more severe than either parent (Endrizzi et al., 1996). The *NAM* gene may play a major role in the spatial regulation of organ initiation and meristem maintenance. *NAM* RNA is present in a ring around the developing shoot apical meristem in the *Petunia* embryo. *NAM* appears to mark the boundary between the meristem and the primordia.

The *STM* expression pattern and mutant phenotypes have shed new light on the ontogeny of the apical meristem and the cotyledons (Kaplan and Cooke, 1997; Kerstetter and Hake, 1997). Because leaves are generated from SAM, a simple explanation (or a logical temporal process) would seem to be that an embryonic apical meristem would form first and generate the embryonic leaves, the cotyledons, and all future apically derived organs. However, *STM* RNA is limited to a few apical cells of the embryo that do not include the progenitor cells of the cotyledons (Kaplan and Cooke, 1997; Endrizzi et al., 1996). Either *STM* is not required for the formation and activity of the embryonic meristem or the cotyledons are formed by a meristem-independent mechanism. A later formation of the meristem is support-

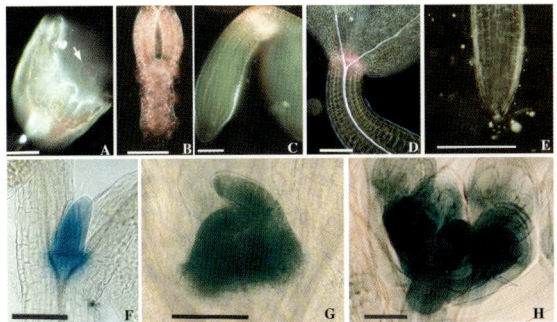

Chapter 3, Fig. 6 *LFY* expression and SAM development. *LFY* promoter-driven GUS activity is seen as a pink stain in A–D and as a blue or green stain in E–H. (A) *LFY*-driven GUS activity as early as globular stage embryo. Arrow points to the embryo proper. (B) GUS activity shown in the entire torpedo stage embryo. (C) GUS activity restricted to shoot apex before desiccation. (D) GUS activity in the shoot apex of a 2-day-old seedling. (E) No GUS activity is detectable in root tips. (F) GUS activity in the shoot apex of a 1-week-old seedling. (G) GUS activity in the shoot apex of a 2-week-old seedling. (H) GUS activity in the influorescence apex. Bars are 75.3 µm.

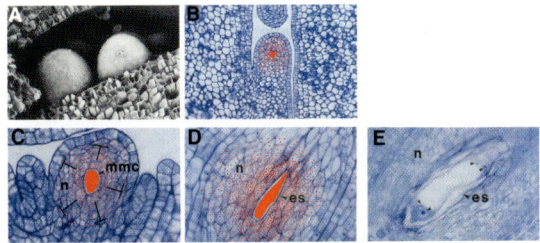

Chapter 3, Fig. 8 Acquisition of embryogenic potential and lateral inhibition of embryo fate determination. Diagrams are redrawn from Koltunow (1993) with a computer-assisted overlay of the embryogenic potential and field shown in red. (A) SEM of *Citrus* ovule primordia. (B) Section through *Citrus* ovule primordium shown in A. (C) *Citrus* ovule showing a megaspore mother cell (mmc). Intensity of red color represents the strength of embryogenic commitment. Bars indicate inhibitory signals from the mmc to suppress embryogenic commitment in cells surrounding the mmc. (D) Young *Citrus* megagametophyte undergoing embryo sac development. (E) *Citrus* embryo sac (es) showing four haploid nuclei (arrowheads) and the nucellus (n).

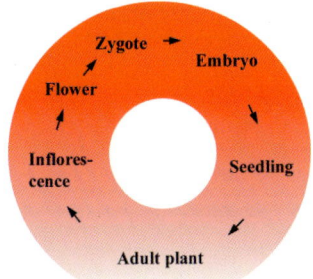

Chapter 3, Fig. 10 Cycling embryogenic fate during the plant life cycle. Color intensity represents the strength of the embryogenic potential. The embryogenic strength of a tissue corresponds to the developmental distance from the zygote (see text). Arrows indicate the direction of developmental progression.

ed by the *stm* mutant phenotype that has normal cotyledons and can also produce axillary shoots with limited leaves (Barton and Poethig, 1993).

D. Mechanism of Cotyledon Formation

Cotyledons form from the tier 1 and half of tier 2 cells of the heart-shaped embryo. The other half of tier 2 cells become part of the hypocotyl. Cell lineage studies showed that tier 1 and tier 2 cells of the heart-shaped embryo originate from different cells of the globular stage *Arabidopsis* embryo (Laux and Jurgens, 1997), which makes it unlikely that some cells were determined to become cotyledons via the segregation of organogenic determinants into different cell lineages of the very early embryo.

During heart-staged development, dicot embryos transition from a spherical to a bipolar structure with bifurcated growth at the shoot end that leads to the formation of the two cotyledons. This morphogenesis requires the normal transport of auxin. Exposure of globular stage embryos to auxin transport inhibitors results in the fusion of the two cotyledons (Liu *et al.*, 1993; Hadfi *et al.*, 1998). Treating monocot embryos with the auxin flow inhibitor naphthyphalamic acid (NPA) shifts the position of the shoot apical meristem relative to the single cotyledon, the scutellum, resulting in the differentiation of duplicated scutella, multiple shoot meristems, or Siamese embryos (Fischer and Neuhaus, 1996; Fischer *et al.*, 1997).

Fused cotyledons or monocotyledons are frequently found in the population of *monopteros (mp)* mutant embryos (hence the name *monopteros*), and *mp* mutants lack both an embryonic axis (hypocotyl and root) and procambium cells. Although *monopteros* mutants do not form an embryonic root, a shoot will germinate and eventually produce an inflorescence. The *mp* inflorescence stem does not transport auxin as efficiently as a wild-type stem (Przemeck *et al.*, 1996). The correlation between impaired auxin transport and aberrant organogenesis is consistent with the notion that auxin transport is involved in axial patterning of the plant embryo.

The *Arabidopsis* mutant *pinformed 1 (pin 1)* also has reduced auxin transport capacity (Okada, 1989; Przemeck *et al.*, 1996) and produces a pinform inflorescence. *PIN 1*, which encodes a transmembrane protein located at the basal end of the xylem parenchyma cells, may function as a carrier for polar auxin transport (Galweiler *et al.*, 1998).

E. A Nongenetic Mechanism

Polarized auxin transport appears to play a major role in the axial patterning processes establishing embryo bisymmetry and organogenesis, as well as procambium differentiation. This raises the questions: How is the auxin flow generated? How do cells know which is the apical and which is the basal end? If the production of auxin were begun preferentially at the shoot end, auxin could flow by diffusion toward the basal cells. What could cause apical cells to produce more auxin than basal cells? Considering that the globular stage embryo is a spherical, apolar structure, a difference could be generated by external signals such as the maternal tissue attached to one end of the embryo. Somatic embryos often have a clump of callus tissue attached to one end that could function like the maternal tissue of the zygote, providing the somatic embryo with a reference point from which to generate an axis. The nonuniform environment surrounding the zygote or the somatic embryo could act as a trigger to initiate embryonic polarity.

Turing's reaction diffusion theory holds that even without an outside trigger, it is inevitable that a physical chemical difference will be generated in a system (Turing, 1952). Living things are in some sense just great big chemical reactions with specific components spread over space and time. Cells, organisms, and biochemistry are all subject to the laws of chemistry and physics. Turing proposes that organisms use the inevitable processes of physical chemistry—diffusion, mass action, and chemical equilibria—to initiate developmental events. The initial event that makes a cell or group of cells different need not be initiated or controlled by macromolecules. These events could be random physicochemical fluctuations that are detected and interpreted by the organism.

Using this theory to model the generation of polar auxin transport, as the plant zygote begins to divide, it assumes a roughly spherical shape, the thermodynamically most stable form. The cells are initially identical with respect to initiating auxin production. The growing embryo is a homogeneous system; the reactants can diffuse freely within the embryo through the plasmodesmata, but homogeneous systems are inherently unstable. They are inevitably disturbed by something internal or external. These random fluctuations create small differences, such as ion currents, that can be amplified and become stabilized into biochemical differences, such as actin microfilament localization or localized auxin synthesis.

This theory does not depend on specific gene action to initiate differences among cells or regions of the embryo. Physicochemical forces drive the differences. A difference is then captured by macromolecules and genes to drive differential gene expression. Physicochemical forces underlie the change from a spherical, radiosymmetrical growth pattern to a growth pattern along an axis. Such a simple biophysical basis of morphogenesis could explain why all plant embryos have the same form despite diverse genetic compositions.

V. Embryo-Specific Genes and Embryonic Mutants

Histogenesis, organogenesis, and meristem activity occur throughout the life of the plant. Many of the genes identified as acting in these processes during embryogenesis also participate in these processes in the adult plant. Consequently, many mutations affecting embryogenesis will also affect organogenesis in the adult plant. The *STM, WUS,* and *ZLL* genes are expressed in all shoot meristems:

embryonic, vegetative, and reproductive. Each of these genes has mutant alleles with a leaky phenotype that allows the growth of seedlings and mutant plants lacking SAMs. These plants show defects in both embryonic and adult development. *PEI1* is necessary only for embryonic apical growth (Li and Thomas, 1998). During embryogenesis, *pei1* embryos failed to form cotyledons. Culture of mutant embryos showed that true leaves could not form, but root formation was apparently normal. Transgenic plants expressing a *PEI1* antisense gene produced white seeds lacking cotyledon, consistent with the mutant phenotype.

These mutant phenotypes suggest the existence of redundant and overlapping pathways regulating histogenesis and organogenesis throughout the life of the plant. These pathways are differentially activated at different stages of the life cycle.

Regulatory pathways and cycles during development can be disrupted by heterochronic mutants that alter the timing of developmental events. Seed formation and germination depend on the correct temporal expression of the many seed development genes expressed in both the embryo and the endosperm. A list of such genes includes some of the seed storage protein genes (Goldberg *et al.*, 1989; Thomas, 1993), the oleosin genes (Vance and Huang, 1988; Hatzopoulos *et al.*, 1990), and the LEA genes (Mundy and Chua, 1988). Genes of the seed maturation program are often regulated by abscisic acid (ABA) and the *ABI3/ VP1* gene (Pla *et al.*, 1991; Giraudat *et al.*, 1992; McCarty *et al.*, 1989, 1991). The *abi3/vp1* mutants do not affect embryo and adult development.

The *LEAFY COTYLEDON 1 (LEC 1)* gene regulates both embryonic fate and the seed maturation program. *LEC 1* expression in postembryonic tissues initiates the formation of embryo-like structures (Lotan *et al.*, 1998). *lec 1* mutants are desiccation intolerant because of the defective expression of some of the seed maturation genes. The shoot meristem is precociously activated in the mutant embryo and trichomes differentiate on the cotyledons. This phenotype suggests premature initiation of postgerminative development in the seed.

The heterochronic mutants *emf* (Chen *et al.*, 1997), *amp1, xtc1, xtc2* (Conway and Poethig, 1997), and *hasty* (Telfer and Poethig, 1998) have no apparent effect on seed maturation. The *amp1, xtc1*, and *xtc2* embryos initiate leaf development prematurely in the seed. The embryos tend to produce enlarged leaf primordia that germinate quickly but form cotyledon-like leaves characterized by a simple venation pattern and the lack of trichomes. Timing of the morphogenetic transitions from globular- to heart- to torpedo- and U-shaped embryo is altered in *xtc1* and *xtc2* mutants, permitting advanced development of the shoot meristem in the mature embryo. The *xtc1* and *xtc2* mutants do not affect postgermination development. Altered timing of meristematic activity in *amp1* mutants results in precocious development of the first leaf with a cotyledon-like phenotype.

Mutations in *EMF* genes display subtle embryonic phenotypes. Weak *emf* mutations affect histological features of the embryonic shoot meristem (Bai and Sung, 1995). Strong mutations also affect the development of the cotyledons

(Chen et al., 1997; Yang et al., 1995). *emf* mutations have their most profound impact in postgermination development. *emf* embryos are precociously reproductive and germinate into reproductive shoot, bypassing vegetative development. The normal function of *EMF* genes may be to suppress reproductive development and promote vegetative development.

This set of heterochronic mutants demonstrates that shoot development is probably regulated by genes of different hierarchal orders. Some genes are specific to the embryo-to-seedling phase transitions. Others, such as *EMF*, are involved in regulating all phases of plant development.

The *pickle* mutant (Ogas et al., 1997) does not display an embryonic phenotype; however, it is unique, being the first gene identified to be involved in the GA-mediated regulation of embryonic/adult root developmental transition. In the absence of GA, mutant roots germinate but develop abnormally and retain embryonic traits.

VI. Embryonic Induction

A. Zygotic Embryogenesis

Zygotes are committed to embryogenic fate. Zygotes always develop into embryos, never leaf or root or an adult plant directly. Embryogenic fate must be induced prior to the formation of the zygote in the ovary tissue that will give rise to the egg, at meiosis, or during oogenesis. It is commonly believed that meiosis resets the genome such that all developmental programs are erased and genome activity is returned to the ground or embryonic state (Ronchi et al., 1992a,b). Whether meiosis in plants is sufficient to make the megaspore embryogenic is unknown. Whatever the process in normal oogenesis, cells other than the egg are or can become embryogenic. Embryos can develop from plant tissues other than the egg (see later) and from callus cultures *in vitro*.

Cell cycle block removal may be the stimulus for initiating embryogeny in the quiescent plant egg. The egg is mitotically arrested until fertilization removes the cell cycle block (Kirshner et al., 1985). Fertilization is not the only stimulus that can remove the cell cycle blockage in frog eggs (Spemann, 1988). A change in Ca^{2+} concentration or a physical stimulus, such as pricking, can initiate embryogenesis. As described later, embryogenic callus lacks this cell division block and develops spontaneously into somatic embryos.

B. Adventive Embryogenesis From Ovary Tissue

Occasionally, embryos develop from cells other than the egg, most commonly from ovary tissue, a process called adventive embryogenesis. In *Citrus* and other

3. Mechanisms of Plant Embryo Development

plant species, nucellar or integument cells can undergo embryogenesis. Successful development of such asexual embryos into seeds is called apomixis (Johri, 1984). Haploid gametophytic cells such as the synergids can develop into haploid embryos (Kasha, 1974). Even cultured microspores can differentiate into haploid embryos. One interpretation of the ability of non-egg cells to undergo embryogenesis is that these gametophytic cells are weakly canalized toward the embryogenic fate. In apomictic species, cells in the vicinity of the megasporocyte develop into diploid embryos directly, bypassing meiosis (Fig. 7; Koltunow, 1993). Cells in the developing ovule appear to be embryogenically competent but not determined to form an embryo.

The concept of "morphogenetic field" derived from animal experimental embryology (for review, see Robertis *et al.*, 1991) can explain adventive embryogenesis. When the central disc of cells that normally give rise to the newt forelimb is removed from the embryo, the surrounding mesodermal cells can produce a forelimb. This is explained as "around the limb-forming cells there is a zone of tissue which has the power, in gradually diminishing intensity . . . to form a limb" (Harrison, 1918). One might envision a zone of tissue in the ovule primordia that has embryogenic potential in gradually diminishing intensity from

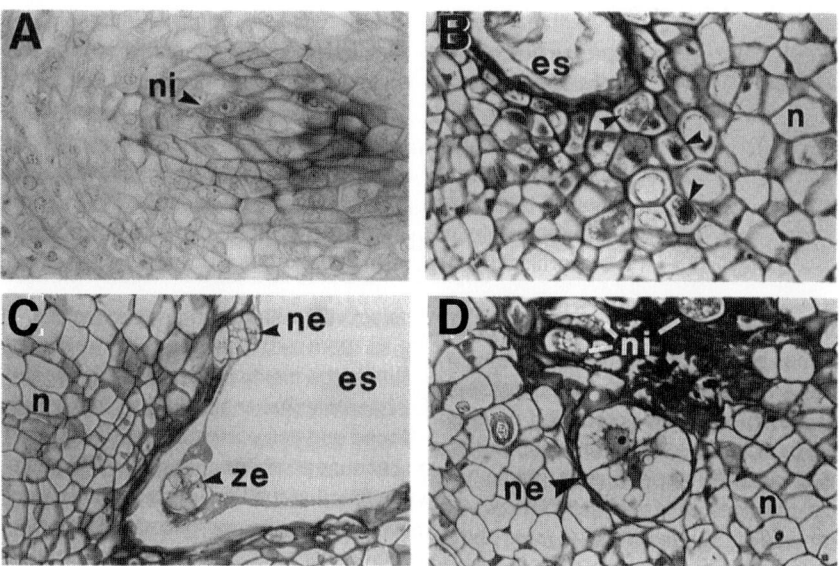

Fig. 7 Adventive embryony in *Valencia*. (A) A nucellar initial (ni) in a fertilized ovule. (B) Nucellar initials (arrowheads) surrounding a fertilized embryo sac (es). (C) A fertilized seed showing the zygotic embryo (ze) and a small nucellar embryo (ne). (D) Nucellar embryo development in unfertilized ovules. A nucellar initial cell has divided to produce a two-celled nucellar embryo. Other nucellar initials are developing in the nucellus (n). (From Koltunow (1993) by permission of the American Society of Plant Physiologists.)

the megasporocyte toward the edge of the primordium (Fig. 8). As a result, most cells in the developing nucellus tissue acquire embryogenic competence, but only the megasporocyte becomes embryogenically committed. As the megasporocyte becomes embryogenically determined, it might inhibit the embryogenic progression in the surrounding cells (Fig. 7), which are subsequently channeled into nucellar or integument development.

The embryogenic fate of the megasporocyte would be propagated to its haploid progeny, but only the egg cell would normally remain strongly canalized toward the embryogenic pathway and become irreversibly committed to embryogenesis. Other cells on the fringe of the morphogenetic field are diverted into pathways that result in the formation of synergids, central cell, antipodal cells, and nucellar cells. In apomictic plants, cells in the vicinity of the egg may also become embryogenically determined, resulting in apogamety, in which cells other than egg in an embryo sac commit to embryogenesis, or adventive embryony, in which nucellar or integument cells commit to embryogenesis (Fig. 7, C and D).

The phenotypes of several suspensor mutants of *Arabidopsis* indicate that suspensor cells are capable of undergoing embryogenesis, if zygotic embryo development is disrupted by one of these mutations (Schwartz *et al.*, 1994). Both zygote daughter cells appear to be embryogenic, but development of the apical cell into an embryo inhibits and changes the embryogenic fate of the basal cell. This could involve cell–cell interactions similar to the lateral signaling during vulval development in *Caenorhabditis elegans* (Kenyon, 1995).

C. Somatic Embryogenesis from Callus

When cultured on synthetic medium, explants of many plant species will dedifferentiate and proliferate into cells that can develop into somatic embryos (Barcelo *et al.*, 1991; Steward *et al.*, 1958). Callus formation is usually induced by culturing explants with 2,4-dichlorophenoxyacetic acid (2,4-D). Removal of 2,4-D results in embryo formation from the callus (Fig. 9). No additional hormones or other stimuli are needed to induce embryo development. It appears that the proliferating cell mass, called embryogenic callus, is already embryogenic.

2,4-D may induce the embryogenic fate directly (Nomura and Komamine, 1986) or indirectly as a result of 2,4-D-stimulated cell division (Goupil *et al.*, 1992). Indirect induction would follow from the 2,4-D-induced cell division, e.g., if the proliferating cells "dedifferentiate" and the epigenetic state of the genome somehow finds itself in the ground or embryogenic state. Once induced, the embryogenic cells would be able to develop into embryos directly, as cell division is not blocked in callus tissue. It is a common misconception that somatic embryos are "induced" from the callus. If one accepts that the embryo-producing callus is already embryogenic, there is no change in the developmental fate when somatic embryos develop from callus cells and thus no induction. Con-

3. Mechanisms of Plant Embryo Development

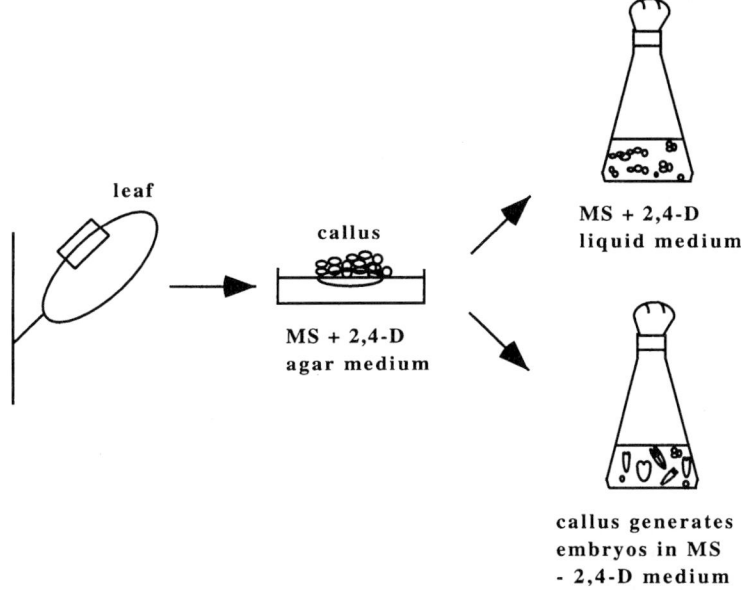

Fig. 9 Somatic embryogenesis of carrot cells. Leaf explants cultured on Murashige and Skoog (MS) medium containing 2,4-D generate callus. When cultures are transferred to 2,4-D-free MS medium, embryos grow from the callus.

sistent with this notion, initiation of somatic embryogeny requires no "inducer," merely removal of the 2,4-D.

The removal of 2,4-D is necessary to permit embryogenesis because 2,4-D not only stimulates cell proliferation, but also prevents the development of somatic embryos on the embryogenic callus. 2,4-D disrupts cellular organization through the stimulation of random cell enlargement and interference with polar transport of the endogenous auxin of the developing embryos. In the presence of 2,4-D, developing embryos become deformed and proliferate into callus again. Some carrot cell lines can form globular stage embryos in 2,4-D but can progress no further (Sung and Okimoto, 1981). The callus remains embryogenic, despite numerous rounds of cell division. If the embryogenic fate is lost, e.g., in cultured single cells, 2,4-D can induce callus growth and the embryogenic fate in these cells again (Nomura and Komamine, 1985, 1986). The embryogenic fate of some cultures can be maintained for many months or years.

Embryogenic callus can be initiated from almost all carrot organs and tissues. Embryogenic callus can be readily generated only from zygotic embryos and immature inflorescence of maize (Pareddy and Petolino, 1990; Tomes and Smith, 1985), barley (Luhrs and Lorz, 1987), wheat (Barcelo et al., 1991), and rice (Cheng and Zhou, 1988). Pine cotyledons are also a good source (Jain et al.,

1989), as are seedling tissues of cauliflower (Pareek and Chandra, 1978) and cotton (Price and Smith, 1979). Mature leaf and stem of these plants are not good sources of embryogenic callus. The varying ability to produce embryogenic callus might reflect the strength of the embryogenic potential in various organs produced at different developmental stages. In general, the closer an organ developmentally to the egg, the more likely it is to be strongly canalized toward embryogenesis (Fig. 10).

D. Changing Embryogenic Competence during the Plant Life Cycle

1. The Decline in Embryogenic Competence during Vegetative Development

The egg or zygote is strongly canalized toward embryogenesis; it always develops into an embryo. The zygotic embryo does not usually produce additional embryos, but when cultured, multiple embryos are seen to bud off the surface, indicating that the epidermal cells are still embryogenic. This has been observed in many species, including *Arabidopsis* (Luo and Koop, 1997). Indeed, embryos are by far the best source of embryogenic callus. Embryogenic competence, as measured by the ease of embryogenic callus production in isolation, decreases as the egg grows into the embryo, the seedling, and finally the vegetative adult (embryogenic competence: egg > embryo > seedling > vegetative shoot).

E. Acquisition of Embryonic Competence during Reproductive Growth

Parts of the plant appear to acquire embryogenic competence long before the egg is formed. Immature inflorescence is often an easy source of embryogenic callus (Barcelo *et al.*, 1991), as is unfertilized ovular tissue (Kochba and Spiegel-Roy, 1973). These observations suggest that the switch from the vegetative to the reproductive state has enabled the tissue to acquire the embryogenic fate. However, in most cases, cells must rid themselves of their cell type-specific "differentiation program" as epidermal or cortical cells of the inflorescence by undergoing dedifferentiation before they can express the embryogenic fate. The dedifferentiation process may require one or more rounds of cell division. Thus somatic embryos are seen to grow either directly from an epidermal cell or indirectly from the explant-generated callus tissue.

In summary, embryogenic competence varies during the plant life cycle. It appears to correspond with the developmental distance of the cell or tissue from the egg. Using the ease of embryogenic callus production as a measure of the embryogenic competence or strength of embryonic potential, one sees a correlation between the embryogenic strength of the tissue and its developmental distance from the zygote (Fig. 10). That a differentiated tissue such as the inflorescence shoot is simultaneously embryogenetically competent indicates that organogenic fate is superimposed over the differentiation program of the tissue and the cell. It would be fascinating to understand the mechanism by which plant cells impart different levels of developmental information.

VII. Summary

1. Evolution in plants has favored both a simpler body plan with fewer cell types and the epigenetic flexibility to regenerate, via growth, dedifferentiation, and redifferentiation, to recover from environmental insults. It has become increasingly apparent that a plant cell uses external signals to differentiate and to maintain or to change the differentiated state. A cell–cell signaling and positional information strategy seems to be the predominant mechanism employed in plant development.

2. An axis can be initiated by physical/chemical forces such as light and ion current, requiring no new gene action. Random chemical fluctuations and physicochemical forces could explain the initiation of differences among cells of equal developmental potential. Amplification of chemical polarizing events may lead to biochemical differences, new gene expression, and finally shoot/root axis establishment.

3. Radial and axial patterning may be governed by a mechanism involving polar auxin transport.

4. Because the meristems and the three fundamental tissues formed during embryogenesis are renewed and extended throughout the life of the plant, with some exceptions, most genes expressed in the embryo are also expressed during postgermination development.

5. Embryogenic competence is acquired during reproductive development. While the zygote is determined for embryogenesis, the developing embryo and often the seedling remain embryogenic. Embryogenic potential declines during vegetative development. The embryogenic strength of a tissue is correlated with its developmental distance from the zygote.

References

Bai, S., and Sung, Z. R. (1995). The role of *EMF1* in regulating the vegetative and reproductive transition in *Arabidopsis thaliana* (Brassicaceae). *Am. J. Bot.* **82**(9), 1095–1103.

Baima, S., Nobili, F., Sessa, G., Lucchetti, S., Ruberti, I., and Morelli, G. (1995). The expression of the *Athb-8* homeobox gene is restricted to provascular cells in *Arabidopsis thaliana*. *Development* **121**(12), 4171–4182.

Barcelo, P., Lazzeri, P. A., Martin, A., and Lorz, H. (1991). Competence of cereal leaf cells. 1. Patterns of proliferation and regeneration capability in vitro of the inflorescence sheath leaves of barley, wheat, and tritordeum. *Plant Sci.* **77**, 243–251.

Barton, M. K., and Poethig, R. S. (1993). Formation of the shoot apical meristem in *Arabidopsis thaliana:* An analysis of development in the wild type and in the *shoot meristemless* mutant. *Development* **119,** 823–831.

Bewley, J. D., and Black, M. (1994). "Seeds: Physiology of Development and Germination." Plenum Press, New York.

Blazquez, M. A., Soowal, L. N., Lee, I., and Weigel, D. (1997). *LEAFY* expression and flower initiation in *Arabidopsis*. *Development* **124,** 3835–3844.

Borkird, C., Choi, J. H., Jin, Z. H., Franz, G., Hatzopoulos, P. Chorneau, R., Bonas, U., Pelegri, F., and Sung, Z. R. (1988). Developmental regulation of embryonic genes in plants. *Proc. Natl. Acad. Sci. USA* **85,** 6399–6403.

Bruck, D. K., and Walker, D. B. (1985). Cell determination during embryogenesis in *Citrus Jambhiri*. I. Ontogeny of the epidermis. *Bot. Gaz.* **146**(2), 188–195.

Camus, G. (1949). Recherches sur le role des bourgeons dans les phenomenes de morphogenese. *Rev. Cytol. Biol. Veg.* **9,** 1.

Chen, L., Cheng, J.-C., Castle, L., and Sung, Z. R. (1997). *EMF* genes regulate *Arabidopsis* inflorescence development. *Plant Cell* **9,** 2011–2024.

Cheng, J. C., Seeley, K. A., Goupil, P., and Sung, Z. R. (1996). Expression of DC8 is associated with, but not dependent on embryogenesis. *Plant Mol. Biol.* **31,** 127–141.

Cheng, J. C., and Zhou, J. Y. (1988). Somatic embryogenesis and plant regeneration from inflorescence of paddy rice. *Am. J. Bot.* **75**(6), 5.

Cheng, J., Seeley, K., and Sung, Z. R. (1995). *RML1* and *RML2*, *Arabidopsis* genes required for cell proliferation at the root tip. *Plant Physiol.* **107,** 365–376.

Chuck, G., Lincoln, C., and Hake, S. (1996). *KNAT1* induces lobed leaves with ectopic meristem when overexpressed in *Arabidopsis*. *Plant Cell* **8,** 1277–1289.

Clark, S. E. (1997). Organ formation at the vegetative shoot meristem. *Plant Cell* **9,** 1067–1076.

Clark, S. E., Jacobsen, S. E., Levin, J., and Meyerowitz, E. M. (1996). The *CLAVATA* and *SHOOT MERISTEMLESS* loci competitively regulate meristem activity in *Arabidopsis*. *Development* **122,** 1567–1575.

Clark, S. E., Williams, R. W., and Meyerowritz, E. M. (1997). The *CLAVATA1* gene encodes a putative receptor kinase that controls shoot and floral meristem size in *Arabidopsis*. *Cell* **89,** 575–585.

Conway, L. J., and Poethig, R. S. (1997). Mutations of *Arabidopsis thaliana* that transform leaves into cotyledons. *Proc. Natl. Acad. Sci. USA* **94,** 10209–10214.

Cutter, E. G. (1969–1971). "Plant Anatomy: Experiment and Interpretation." Edward Arnold, London.

Dolan, L., Janmaat, K., Willemsen, V., Linstead, P., Poethig, S., Roberts, K., and Scheres, B. (1993). Cellular organisation of the *Arabidopsis thaliana* root. *Development* **119,** 71–84.

Endrizzi, K., Moussian, B., Haecker, A., Levin, J., and Laux, T. (1996). The *SHOOT MERISTEMLESS* gene is required for maintenance of undifferentiated cells in *Arabidopsis* shoot and floral meristems and acts at a different regulatory level than the meristem genes *WUSCHEL* and *ZWILLE*. *Plant J.* **10,** 967–979.

Evans, M. M. S., and Barton, M. K. (1997). Genetics of angiosperm shoot apical meristem development. *Annu. Rev. Plant Physiol. Plant Mol. Biol.* **48,** 673–701.

Fischer, C., and Neuhaus, G. (1996). Influence of auxin on the establishment of bilateral symmetry in monocots. *Plant J.* **9,** 659–669.

Fischer, C., Speth, V., Fleig-Eberenz, S,. and Neuhaus, G. (1997). Induction of zygotic polyembryos in wheat: Influence of auxin polar transport. *Plant Cell* **9,** 1767–1780.

Fry, S. C., and Wangermann, E. (1976). Polar transport of auxin through embryos. *New Phytol.* **77,** 313–317.

Galweiler, L., Guan, C., Muller, A., Wisman, E., Mendgen, K., Yephremov, A., and Palme, K. (1998). Regulation of polar auxin transport by AtPIN1 in *Arabidopsis* vascular tissue. *Science* **282**(5397), 2226–2230.

3. Mechanisms of Plant Embryo Development

Giraudat, J., Hauge, B. M., Valon, C., Smalle, J., Parcy, F., and Goodman, H. M. (1992). Isolation of the *Arabidopsis* ABI3 gene by positional cloning. *Plant Cell* **4,** 1251–1261.

Goldberg, R. B., Barker, S. J., and Perez-Grau, L. (1989). Regulation of gene expression during plant embryogenesis. *Cell* **56,** 149–160.

Goupil, P., Hatzopoulos, P., Franz, G., Hempel, F. D., You, R., and Sung, Z. R. (1992). Transcriptional regulation of a seed-specific carrot gene, DC8. *Plant Mol. Biol.* **18,** 1049–1063.

Green, P. B. (1992). Pattern formation in shoots: A likely role for minimal energy configurations of the tunica. *Int. J. Plant Sci.* **153**(3), S59–S75.

Grossniklaus, U., Viell-Calzada, J.-P., Howppner, M. A., and Gagliano, W. B. (1998). Maternal control of embryogenesis by *MEDEA*, a polycomb group gene in *Arabidopsis*. *Science* **280,** 446–450.

Hadfi, K., Speth, V., and Neuhaus, G. (1998). Auxin-induced developmental patterns in *Brassica juncea* embryos. *Development* **125,** 879–887.

Hardtke, C. S., and Berleth, T. (1998). The *Arabidopsis* gene *MONOPTEROS* encodes a transcription factor mediating embryo axis formation and vascular development. *EMBO J.* **17**(5), 1405–1411.

Harrison, R. G. (1918). Experiments on the development of the fore-limb of Amblystoma, a self-differentiating equipotential system. *J. Exp. Zoo.* **25,** 413–461.

Hatzopoulos, P., Franz, G., Choy, L., and Sung, Z. R. (1990). Interaction of nuclear factors with upstream sequences of a lipid body membrane protein gene from carrot. *Plant Cell* **2,** 457–467.

Jackson, D., Veit, B., and Hake, S. (1994). Expression of maize *KNOTTED1* related homeobox genes in the shoot apical meristem patterns of morphogenesis in the vegetative shoot. *Development* **120,** 405–413.

Jain, S. M., Dong, N., and Newton, R. J. (1989). Somatic embryogenesis in slash pine (*Pinus elliottii*) from immature embryos cultures *in vitro*. *Plant Sci.* **65,** 233–241.

Jeong and Clark. (1997). Histological and molecular description of vascular tissue development in shoot apical meristem mutants in *Arabidopsis*. *In* "8th International Conference on *Arabidopsis* Research," Madison, WI.

Johansen, D. A. (1950). "Plant Embryology." Chronica Botanica, Waltham, MA.

Johri, B. M. (1984). "Embryology of Angiosperms." Springer-Verlag, Berlin.

Kaplan, D. R. (1984). The concept of homology and its central role in the elucidation of plant systematic relationships. *In* "Cladistics: Perspectives on the Reconstruction of Evolutionary History" (T. Duncan and T. F. Stuessy, eds.), pp. 51–70. Columbia Univ. Press, New York.

Kaplan, D. R., and Cooke, T. J. (1997). Fundamental concepts in the embryogenesis of dicotyledons: A morphological interpretation of embryo mutants. *Plant Cell* **9,** 1903–1919.

Kaplan, D. R., and Hagemann, W. (1992). The organism and plant cells in light of Goethe's comparative morphological method. *In* "der Mitte Zwischen Natur und Subjekt. Johann Wolfgang Goethes Versuch, die Metamorphosen der Pflanzen zu erklaren 1790–1990. Sachverhalte, Gedanken, Wirkungen" (G. Mann, D. Mollenhauer, and S. Peters, eds.), pp. 93–117. Waldemar Kramer, Frankfurt, Germany.

Kasha, K. (1974). "Haploids in Higher Plants: Advances and Potential." University of Guelph, Guelph.

Kemp, M. (1943). Morphological and ontogenetic studies on *Torreya californica* Torr. I. The vegetative apex of the megasporangiate tree. *Am. J. Bot.* **30**(7), 504–517.

Kenyon, C. (1995). A perfect vulva every time: Gradients and signaling cascades in *C. elegans*. *Cell* **82,** 171–174.

Kerstetter, R. A., and Hake, S. (1997). Shoot meristem formation in vegetative development. *Plant Cell* **9,** 1001–1010.

Kirshner, M., Newport, J., and Gerhart, J. (1985). The timing of early developmental events in *Xenopus*. *Trends Genet.* Feb. 41–47.

Kochba, J., and Spiegel-Roy, P. (1973). Effect of culture media on embryoid formation from ovular callus of 'Shamouti' orange (*Citrus sinensis*) *Z. Pflanzenzucht.* **69,** 156–162.

Koltunow, A. M. (1993). Apomixis: Embryo sacs and embryos formed without meiosis or fertilization in ovules. *Plant Cell* **5,** 1425–1437.

Konar, R. N., and Nataraja, K. (1965). Experimental studies in *Ranunculus sceleratus* L. Development of embryos from the stem epidermis. *Phytomorphology* **15,** 132–137.

Koornneef, M., Nanhart, C. J., Hilhorst, H. W. M., and Karseen, C. M. (1989). In vitro inhibition of seed development and reserve protein accumulation in recombinants of abscisic acid biosynthesis and responsiveness mutants in *Arabidopsis thaliana. Plant Physiol.* **90,** 463–469.

Laux, T., and Jurgens, G. (1997). Embryogenesis: A new start in life. *Plant Cell* **9,** 989–1000.

Laux, T., Mayer, K. F. X., Berger, J., and Jurgens, G. (1996). The *WUSCHEL* gene is required for shoot and floral meristem integrity in *Arabidopsis. Development* **122,** 87–96.

Li, Z., and Thomas, T. L. (1998). *PEI1*, an embryo-specific zinc finger protein gene required for heart-stage embryo formation in *Arabidopsis. Plant Cell* **10**(3), 383–399.

Liu, C. M., Zu, Z. H., and Chua, N. H. (1993). Auxin polar transport is essential for the establishment of bilateral symmetry during early plant embryogenesis. *Plant Cell* **5,** 621–630.

Lolle, S. J., Hsu, W., and Pruitt, R. E. (1998). Genetic analysis of organ fusion in *Arabidopsis thaliana. Genetics* **149,** 607–619.

Long, J. A., Moan, E. I., Medford, J. L., and Barton, M. K. (1996). The *Arabidopsis SHOOTMERISTEMLESS* gene encodes a member of the *KNOTTED* class of homeodomain proteins. *Nature* **379,** 66–69.

Lotan, T., Ohto, M., Yee, K. M., West, M. A., Lo, R., Kwong, R. W., Yamagishi, K., Fischer, R. L., Goldberg, R. B,. and Harada, J. J. (1998). *Arabidopsis LEAFY COTYLEDON1* is sufficient to induce embryo development in vegetative cells. *Cell* **93**(7), 1195–1205.

Lu, P., Porat, R., Nadeau, J. A., and O'Neill, S. D. (1996). Identification of a meristem L1 layer-specific gene in *Arabidopsis* that is expressed during embryonic pattern formation and defines a new class of homeobox genes. *Plant Cell* **8,** 2155–2168.

Luhrs, R., and Lorz, H. (1987). Plant regeneration in vitro from embryogenic cultures of spring- and winter-type barley (*Hordeum vulgare* L.) varieties. *Theor. Appl. Genet.* **75,** 16–25.

Luo, Y., and Koop, H. U. (1997). Somatic embryogenesis in cultured immature zygotic embryos and leaf protoplasts of *Arabidopsis thaliana* ecotypes. *Planta* **202**(3), 387–396.

Lyndon, R. F. (1990). "Plant Development: The Cellular Basis." Unwin Hyman, Boston, MA.

Mattsson, J., Sung, Z. R., and Berleth, T. (1999). Responses of plant vascular systems to auxin transport inhibition. *Development* **126**(13), 2979–2991.

Mayer, K. F., Schoof, H., Haecker, A., Lenhard, M., Jurgens, G., Laux, T. (1998). Role of *WUSCHEL* in regulating stem cell fate in the *Arabidopsis* shoot meristem. *Cell* **95**(6), 805–815.

McCarty, D. R., Carson, C. B., Stinard, P. S., and Robertson, D. S. (1989). Molecular analysis of *viviparous-1*: An abscisic acid-insensitive mutant of maize. *Plant Cell* **1,** 523–532.

McCarty, D. R., Hattori, T., Carson, C. B., Vasil, V., Lazon, M., and Vasil, I. K. (1991). The *viviparous-1* developmental gene of maize encodes a novel transcriptional activator. *Cell* **66,** 895–905.

Medford, J. I. (1992). Vegetative apical meristems. *Plant Cell* **4,** 1029–1039.

Miksche, J. P., and Brown, J. A. M. (1965). Development of vegetative and floral meristems of *Arabidopsis thaliana. Am. J. Bot.* **52**(6), 533–537.

Miksche, J. P., and Brown, J. A. M. (1965). Development of vegetative and floral meristems of *Arabidopsis thaliana. Am. J. Bot.* **52**(6), 533–537.

Moussian, B., Schoof, H., Haecker, A., Jurgens, G., and Laux, T. (1998). Role of the *ZWILLE* gene in the regulation of central shoot meristem cell fate during *Arabidopsis* embryogenesis. *EMBO J.* **17**(6), 1799–1809.

Mundy, J., and Chua, N.-H. (1988). Abscisic acid and water-stress induce the expression of a novel rice gene. *EMBO J.* **7**(8), 2279–2286.

Nomura, K., and Komamine, A. (1985). Identification and isolation of single cells that produce somatic embryos at a high frequency in a carrot suspension culture. *Plant Physiol.* **79,** 988–991.

Nomura, K., and Komamine, A. (1986). Polarized DNA synthesis and cell division in cell clusters during somatic embryogenesis from single carrot cells. *New Phytol.* **104,** 25–32.

3. Mechanisms of Plant Embryo Development

Ogas, J., Cheng, J. C., Sung, Z. R., and Somerville, C. (1997). Cellular differentiation regulated by gibberellin in the *Arabidopsis thaliana pickle* mutant. *Science* **277,** 91–94.

Okada, K., Komaki, M. K., and Shimura, Y. (1989). Mutational analysis of pistil structure and development of *Arabidopsis thaliana*. *Cell Diff. Dev.* **28,** 27–38.

Pareddy, D. R., and Petolino, J. F. (1990). Somatic embryogenesis and plant regeneration from immature inflorescences of several elite inbreds of maize. *Plant Sci.* **67,** N2, 211–219.

Pareek, L. K., and Chandra, N. (1978). Somatic embryogenesis in leaf callus from cauliflower (*Brassica oleracea* var. *botrytis*). *Plant. Sci. Lett.* **11**(3/4), 311–316.

Pla, M., Gomez, J., Goday, A., and Pages, M. (1991). Regulation of the abscisic acid-responsive gene *rab28* in maize *viviparous* mutants. *Mol. Gen. Genet.* **230,** 394–400.

Poddubnaya-arnoldi (1967). Comparative embryology of the Orchidaceae. *Phytomorphology* **17,** 312–320.

Poethig, R. S. (1987). Clonal analysis of cell lineage patterns in plant development. *Am. J. Bot.* **74**(4), 581–594.

Poethig, R. S., Coe, E. H., Jr., and Johri, M. M. (1986). Cell lineage patterns in maize embryogenesis: A clonal analysis. *Dev. Biol.* **117,** 392–404.

Pollock, E. G., and Jensen, W. A. (1964). Cell development during early embryogenesis in *Capsella* and Gossypium. *Am. J. Bot.* **51,** 915–921.

Price, H. J., and Smith, R. H. (1979). Somatic embryogenesis in suspension cultures of *Gossypium klotzschianum* Anderss. *Planta* **145,** 305–307.

Przemeck, G. K., Mattsson, J., Hardtke, C. S., Sung, Z. R., and Berleth, T. (1996). Studies on the role of the *Arabidopsis* gene *MONOPTEROS* in vascular development and plant cell axialization. *Planta* **200**(2), 229–237.

Rao, A. N. (1967). Flower and seed development in *Arundina graminifolia*. *Phytomorphology* **17,** 291–300.

Ray, S., Golden, T., and Ray, A. (1996). Maternal effects of the shot integument mutation on embryo development in *Arabidopsis*. *Dev. Biol.* **180**(2), 365–369.

Robertis, E. M., Morita, E. A., and Cho, K. W. Y. (1991). Gradient fields and homeobox 1genes. *Development* **112,** 669–678.

Ronchi, V. N., Giorgetti, L., Tonelli, M. G., and Martini, G. (1992a). Ploidy reduction and genome segregation in cultured carrot cell lines. I. Prophase chromosome reduction. *Plant Cell Tissue Organ Culture* **30,** 107-114.

Ronchi, V. N., Giorgetti, L., Tonelli, M. G., and Martini, G. (1992b). Ploidy reduction and genome segregation in cultured carrot cell lines. II. Somatic meiosis. *Plant Cell Tissue Organ Culture* **30,** 115–119.

Sachs, T. (1991). Cell polarity and tissue patterning in plants. *Development* Supp. (1), 83–93.

Schiavone, F. M., and Racusen, R. H. (1991). Regeneration of the root pole in surgically transected carrot embryos occurs by position-dependent, proximodistal replacement of missing tissues. *Development* **113,** 1305–1313.

Schulz, R., and Jensen, W. A. (1968). *Capsella* embryogenesis: The egg, zygote, and young embryo. *Am. J. Bot.* **55**(7), 807–819.

Schwartz, B. W., Yeung, E. C., and Meinke, D. W. (1994). Disruption of morphogenesis and transformation of the suspensor in abnormal suspensor mutants of *Arabidopsis*. *Development* **120,** 3235–3245.

Sinha, N. R., Williams, R. E., and Hake, S. (1993). Overexpression of the maize homeo box gene, *KNOTTED1*, causes a switch from determinate to indeterminate cell fates. *Genes Dev.* **7,** 787–795.

Smith, L. G., Jackson, D., and Hake, S. (1995). Expression of *knotted1* marks shoot meristem formation during maize development. *Dev. Genet.* **16,** 344–348.

Souer, E., van Houwelingen, A., Kloos, D., Mol, J., and Koes, R. (1996). The *NO APICAL MERISTEM* gene of *Petunia* is required for pattern formation in embryos and flowers and is expressed at meristem and primordia boundaries. *Cell* **85,** 159–170.

Spemann, H. (1988). "Embryonic Development and Induction." Garland Publishing, New York.
Steeves, T. A., and Sussex, I. M. (1988). "Patterns in Plant Development," 2nd Ed. Cambridge Univ. Press, Cambridge, New York.
Sterk, P., Booij, H., Schellekens, G. A., van Kammen, A., and de Vries, S. C. (1991). Cell-specific expression of the carrot EP2 lipid transfer protein gene. *Plant Cell* **3,** 907–921.
Stewart, F. C., Mapes, M. O., and Mears, D. (1958). Growth and organized development of cultured cells. II. Organization in cultures grown from freely suspended cells. *Am. J. Bot.* **45,** 705–707.
Sung, Z. R., and Okimoto, R. (1981). Embryonic proteins in somatic embryos of carrot. *Proc. Natl. Acad. Sci. USA* **78,** 3683–3687.
Telfer, A., and Poethig, R. S. (1998). *HASTY*: A gene that regulates the timing of shoot maturation in *Arabidopsis thaliana*. *Development* **125**(10), 1889–1898.
Thomas, T. L. (1993). Gene expression during plant embryogenesis and germination: An overview. *Plant Cell* **5**(10), 1401–1410.
Tomes, D. T., and Smith, O. S. (1985). The effect of parental genotype on initiation of embryogenic callus from elite maize (*Zea mays* L.) germplasm. *Theor. Appl. Genet.* **70,** 505–509.
Toonen, M. A., Verhees, J. A., Schmidt, E. D., van Kammen, A., and de Vries, S. C. (1997). AtLTP1 luciferase expression during carrot somatic embryogenesis. *Plant J.* **12**(5), 1213–1221.
Turing, A. M. (1952). The chemical basis of morphogenesis. *Phil. Trans. Roy. Soc. Lond.* **237,** 37–72.
Ulmasov, T., Hagen, G., and Guilfoyle, T. J. (1997). ARF1, a transcription factor that binds to auxin response elements. *Science* **276,** 1865–1868.
van den Berg, C., Willemsen, V., Hendriks, G., Weisbeek, P., and Scheres, B. (1997). Short-range control of cell differentiation in the *Arabidopsis* root meristem. *Nature* **390**(20 Nov.), 287–289.
Vernon, D. M., and Meinke, D. W. (1994). Embryogenic transformation of the suspensor in *twin*, a polyembryonic mutant of *Arabidopsis*. *Dev. Biol.* **165,** 566–573.
Vance, V. B., and Huang, A. H. C. (1988). Expression of lipid body protein gene during maize seed development. *J. Biol. Chem.* **263**(3), 1476–1481.
Wardlaw, C. W. (1953). "Embryogenesis in Plants." Methuen & Co., London.
Wardlaw, C. W. (1953). Comparative observations on the shoot apices of vascular plants. *New Phytol.* 53.
Wardlaw, C. W. (1953). A commentary on Turing's diffusion reaction theory of morphogenesis. *New Phytol.* **52,** 40.
Yadegari, R. G., de Paiva, R., Laux, T., Koltunow, A. M., Apuya, N., Zimmerman, J. L., Fischer, R. L., Harada, J. J., and Goldberg, R. B. (1994). Cell differentiation and morphogenesis are uncoupled in *Arabidopsis raspberry* embryos. *Plant Cell* **6**(12), 1713–1729.
Yang, C.-H., Chen, L., and Sung, Z. R. (1995). Genetic regulation of shoot development in *Arabidopsis*: Role of the EMF genes. *Dev. Biol.* **169,** 421–435.
Yeung, E. C., and Clutter, M. E. (1978). Embryogeny of *Phaseolus coccineus*: Growth and microanatomy. *Protoplasm* **94,** 19–40.
You, R., and Jensen, W. A. (1985). Ultrastructural observations of the mature megagametophyte and the fertilization in wheat (*Triticum aestivum*). *Can. J. Bot.* **63,** 163–178.
Zhang, J. Z., and Somerville, C. R. (1997). Suspensor-derived polyembryony caused by altered expression of valyl-tRNA synthetase in the *twn2* mutant of *Arabidopsis*. *Proc. Natl. Acad. Sci. USA* **94,** 7349–7355.

4

Sperm-Mediated Gene Transfer

Anthony W. S. Chan,[1,2] C. Marc Luetjens,[3]
and Gerald P. Schatten[1,2,4]
[1]Oregon Regional Primate Research Center
Beaverton, Oregon 97006
[2]Departments of Obstetrics and Gynecology and
[4]Cell and Developmental Biology
Oregon Health Sciences University
Portland, Oregon 97201
[3]Center for Interdisciplinary Clinical Research
Research Group "Programmed Cell Death"
Westphalian Wilhelms-University
D-48149 Müenster, Germany

I. Introduction
II. DNA Binding
III. DNA Internalization
IV. DNA Integration
V. Alternative Strategies
VI. Perspectives
   References

I. Introduction

The first transgenic mouse was created in the early 1990s via incubation of spermatozoa with exogenous DNA followed by *in vitro* fertilization (IVF) (Lavitrano *et al.*, 1989). The success led to a new interest in transgenic technology because of its simplicity, although various methods have been successfully used to create transgenic animals (Gandolfi *et al.*, 1989; Bird *et al.*, 1992; Schellander *et al.*, 1995; Sperandio *et al.*, 1996; Chan, 1998). However, the sperm vector initiated a strong debate on its credibility as many laboratories have been unable to create transgenic animal using sperm vectors (Brinster *et al.*, 1989). Nevertheless, the concept of using sperm to carry exogenous DNA into an oocyte during fertilization was first considered in 1971. Brackett and colleagues (1971) demonstrated the delivery of viral DNA into rabbit oocytes by spermatozoa after the uptake of DNA. The debate has diminished as transgenic animals have been produced in various species using sperm as a DNA carrier. These include mice (Maione *et al.*, 1998), pigs (Gandolfi *et al.*, 1989; Sperandio *et al.*, 1996), fish

(Khoo et al., 1992), cattle (Schellander et al., 1995), chicken (Nakanishi and Iritani, 1993), and *Xenopus* (Kroll and Amaya, 1996). There are three critical steps for successful sperm-mediated gene transfer to create transgenic animals. The first step is the binding of DNA on the sperm cell surface, followed by the internalization of foreign DNA into the nucleus, and finally the integration of exogenous DNA. Although the mechanisms of each individual step are not fully understood, many advancements in the past few years have provided a better understanding of the process. This review targets individual steps, discussing the current problems and possible solutions.

II. DNA Binding

Intensive studies regarding the relationship between exogenous DNA and spermatozoa have demonstrated that the binding of DNA on the sperm cell surface is not a random event (Camaioni et al., 1992; Zani et al., 1995). Proteinase treatment of sperm–DNA complexes abolishes the sperm's binding ability, which suggest the involvement of specific protein interactions between sperm and DNA (Zani et al., 1995; Spadafora, 1998). The addition of polyanions such as heparin inhibits the interaction between exogenous DNA and sperm, which may be explained by the participation of the polyanionic backbone of the DNA (Zani et al., 1995; Macha et al., 1997). Because seminal plasma contains polyamines and glycosaminoglycans, the competition of these molecules with DNA for binding on the sperm cell surface is expected (Camaioni et al., 1992). Binding efficiency gives an indication of the amount of DNA that interacts with sperm, but does not necessarily represent the internalization of foreign DNA into an inner compartment of the sperm head such as the nucleus. Due to the complexity of the sperm nucleus, the question of integration has been difficult to address. It is suggested that the integration event may not take place in the sperm nucleus immediately, but later on during the fertilization process (Chan, 1999). Indeed, the knowledge in terms of when and where this integration occurs is limited.

A tremendous amount of effort has focused on the sperm membrane and nucleus structure. The binding of exogenous DNA has been shown by different laboratories in various species, including boar, bull, ram, buffalo, carp, rabbit, sea urchin, fish, honeybee, blowfly, xenopus, and rooster (Brackett et al., 1971; Camaioni et al., 1992; Khoo et al., 1992; Nakanishi and Iritani, 1993; Habrova et al., 1996; Macha et al., 1997; Spadafora, 1998). However, to date, transgenic animals have only been created in a limited number of species. Once again, this indicates that the binding of DNA on the sperm cell surface is an essential step, but is not the determining factor for successful gene transfer. It has been demonstrated that only a limited portion of sperm binds to DNA, estimated to be between 10 and 80%, and this varies among species (Bachiller et al., 1991; Francolini et al., 1993). Time required for exogenous DNA to bind sperm is suggested between 10 and 40 minutes (Gandolfi et al., 1989; Lavitrano et al., 1992; Lauria and Gandolfi, 1993). The interaction between sperm and DNA is considered to be a reversible reaction as prolonged incubation does not reach 100% binding efficiency (Macha et al., 1997). Macha and colleagues have shown that the amount of DNA bound on sperm increases when both the amount of DNA and the time of incubation increase prior to reaching saturation (Francolini et al., 1993; Macha et al., 1997). Binding efficiency was determined primarily by using radioisotope-labeled DNA fragments, autoradiography, or *in situ* hybridization techniques (Brackett et al., 1971; Camaioni et al., 1992; Lavitrano et al., 1992; Patil and Khoo, 1995). Determination of the DNA-binding efficiency on sperm prior to fertilization has not yet been possible, and the controversy between the binding of DNA on live (Francolini et al., 1993) and dead (Bird et al., 1992) sperm has led to the concerns of the selection or exclusion of DNA bound on the sperm surface during fertilization. So far, no concrete solution has been suggested, yet the answer to this controversial issue may provide an explanation to the inconsistent rate of transgenic animal produced by sperm vectors. Sperm–DNA-binding efficiency has been shown to be highly variable among experiments, species, and individuals (Bachiller et al., 1991; Francolini et al., 1993). Therefore, evaluation of the binding efficiency in live sperm prior to fertilization may help predict the outcome of the rate of transgenesis. The use of rhodamine-tagged DNA has been shown to be a novel way to monitor the dynamics of DNA fragments bound on the sperm cell surface before and after fertilization (Chan et al., 2000). Rhodamine-tagged plasmid DNA binds primarily at the equatorial and postacrosomal regions of sperm, which is consistent with previous studies using radioisotope-labeled DNA (Gandolfi et al., 1989; Camaioni et al., 1992; Francolini et al., 1993; Lauria and Gandolfi, 1993). The use of this live imaging system has shown that sperm bound with higher amounts of DNA seem to have lower motility than those having less DNA bound. This implies that *in vitro* fertilization is more likely with sperm that carry less DNA and may result in a lower transgenic rate.

III. DNA Internalization

The internalization of foreign DNA into sperm nuclei has been debated for many years (Bachiller et al., 1991; Francolini et al., 1993; Lauria and Gandolfi, 1993; Kim et al., 1997; Lavitrano et al., 1997; Macha et al., 1997; Huguet and Esponda, 1998; Spadafora, 1998). The fact that larger DNA fragments are more easier to be taken up by sperm cells than smaller fragments indicates the involvement of an active internalization mechanism (Lauria and Gandolfi, 1993). Translocation of exogenous DNA into sperm nuclei has been reported (Francolini et al., 1993; Lauria and Gandolfi, 1993; Patil and Khoo, 1995; Lavitrano et al., 1997; Macha et al., 1997; Huguet and Esponda, 1998; Magnano et al., 1998; Spadafora, 1998). It was suggested that 15–22% of the bound DNA translocated into the sperm cell nucleus spontaneously (Lavitrano et al., 1992; Francolini et al., 1993). The amount of remaining DNA after DNase I treatment also supports the internalization of foreign DNA into sperm (Atkinson et al., 1991; Macha et al., 1997), perhaps into the inner compartment of the sperm head rather than the nucleus. Nevertheless, Macha and colleagues (1997) reported that an average of 50–70 DNA molecules were translocated into mouse and *Xenopus* sperm nuclei after 20 min incubation. In contrast, Lauria and Gandolfi (1993) suggested that DNA was not internalized but only bound on the surface, as foreign DNA was ultimately diluted and disappeared during preimplantation development. Kim and colleagues (1997) also suggested that DNA translocation into sperm nuclei could be an artifact. Most of the analyses, such as *in situ* hybridization and ultra-thin electronic microscopy, require hyper- and hypoosmolar solution treatments, which might facilitate the translocation of surface-bound DNA into the nucleus (Kim et al., 1997). The hypothesis of receptor-mediated translocation of foreign DNA into sperm cell nuclei has been proposed (Lavitrano et al., 1997; Spadafora, 1998). Indeed, receptor-mediated gene transfer in mouse embryos and the creation of transgenic animals have been demonstrated. This supports the hypothesis of foreign DNA translocation into sperm cell nuclei through membrane receptors (Ivanova et al., 1999). The expression of major histocompatibility complex (MHC) class II genes has been found to be crucial for sperm–DNA interactions, although MHC II molecules are not present on the sperm cell surface (Mori et al., 1990; Wu et al., 1990; Lavitrano et al., 1997; Spadafora, 1998). In MHC class II knockout mice, the binding efficiency of DNA is significantly decreased, which implies its role in sperm–DNA interactions (Spadafora, 1998). Similarly, CD4 molecules on the plasma membrane of the sperm head also play an important role in DNA internalization, as internalization of DNA was inhibited in CD4 knockout mice (Lavitrano et al., 1997; Spadafora, 1998). Spadafora (1998) suggested that exogenous DNA interacts with DNA-binding protein (DBP) and that DNA–BDP nucleoprotein complexes trigger the CD-4 mediated internalization, through the nucleopores, reach the nuclear matrix, and release inside the nucleus. Although this suggested model implicates the active transport of DNA molecules into sperm nuclei, further clarification is needed.

IV. DNA Integration

One of the interesting observations in transgenic mice (Maione et al., 1998; Spadafora, 1998) and other species, including chicken (Rottmann et al., 1991), zebrafish (Khoo et al., 1992), bovine (Schellander et al., 1995; Sperandio et al., 1996), and pig (Sperandio et al., 1996), was the identical insertion site of the transgene. This is consistent with the latest finding of cloning an insertion site in sperm cells after coincubation with plasmid DNA (Zoraqi and Spadafora, 1997; Magnano et al., 1998). There might be several explanations for this uniform integration pattern: (1) integration of exogenous DNA into the embryonic genome may occur at a unique time point when chromatin is accessible, such as during the decondensation of the sperm nucleus or the formation of male and female pronuclei; (2) integration may not occur immediately after sperm–DNA interaction but later during fertilization or oocyte activation; or (3) integration of exogenous DNA may require the aid of maternal machinery, such as a DNA repair mechanism (Brinster *et al.*, 1985; Coffin, 1990). Although all assumptions are in favor of the integration event taking place after sperm penetration, which may also require decondensation of the sperm nucleus and activation of the oocyte. However, integration in sperm nuclei prior to fertilization may occur.

Zoraqi and Spadafora (1997) suggested that the internalization of DNA is closely associated with the nuclear matrix, and extensive rearrangement of the plasmid DNA was observed followed by recombination between exogenous DNA and sperm genomic DNA. In their study, based on the sequence of the cloned insertion site in sperm nuclei, the involvement of topoisomerase II is also considered, as a topoisomerase II consensus sequence was found at one end of the insertion site. In addition, Maione and colleagues (1997) have demonstrated the activation of endonuclease induced by interaction between mature sperm cells and exogenous DNA, perhaps related to the apoptosis of sperm cells. They suggested that exogenous DNA was specifically cleaved upon internalization by sperm endonuclease. Furthermore, the transcriptional and enzymatic activity in mature sperm has been demonstrated (Kramer and Krawetz, 1997; Stewart *et al.*, 1999). Transcription of testis-specific genes, including protamine (*PRM1* and *PRM2*) and transition protein (*TNP1* and *TNP2*), has been demonstrated in round spermatids (Kleene *et al*, 1984; Wykes *et al.*, 1995). In addition to transcriptional activity, the presence of active DNA polymerase in human, bull, and mouse spermatozoa suggests a role in fertilization and embryogenesis (Hecht, 1974; Witkin *et al.*, 1975; Chevaillier and Philippe, 1976) and a possible role in exogenous DNA integration. Further evidence against the theory of a quiescent genome is the somatic-like organization of selected regions in the sperm nucleus (Kramer and Krawetz, 1997). Sperm DNA is organized into loop domains and is attached to the sperm nuclear matrix, bound with protamines (Ward *et al.*, 1999). In addition to the specific organization of the chromatin structure, a unique spatial positioning of the haploid genome has also been suggested (Zalensky *et al.*, 1995; Luetjens *et al.*, 1999). It has been shown that the organization of the loop domain is important in DNA replication and transcriptional regulation (Cody *et al.*, 1993; Barone *et al.*, 1994; Kramer and Krawetz, 1996; Singh *et al.*, 1997). In addition to the relationship with the nuclear matrix, the presence of a DNase I sensitivity site within the haploid genome indicates potential transcription activity (Kramer and Krawetz, 1996). Nevertheless, active replication and transcription regions have been suggested to be the favorable site for DNA insertion (Coffin, 1990; Rijkers *et al.*, 1994; Chan, 1999). After all, the sperm nucleus is not completely inert and DNA integration in sperm nuclei should not be excluded. Indeed, about 15% of the histone protein remains bound on sperm chromatin, together with 85% of protamine (Kramer and Krawetz, 1997). Although most of the sperm haploid genome is tightly bound and is not considered a favorable site for DNA insertion, the existence of a somatic-like region may create a preferential target site for DNA insertion.

Regarding the studies in transgenic mice, identical insertion sites in the host cell genome have raised an interesting question. Are there any preferential or intrinsic factors that direct the integration process? Although the most frequently studied species created by sperm vectors is mice, only a limited number of laboratories can perform the procedure successfully (Lavitrano *et al.*, 1989; Maione *et al.*, 1998; Spadafora, 1998). The analysis of integration sites in transgenic mice and the subcloning of the integrated plasmid DNA demonstrate a uniform integration pattern (Zoraqi and Spadafora, 1997; Magnano *et al.*, 1998; Maione *et al.*, 1998). These results support the idea of a preferential site created by the spatial organization of the haploid genome. However, the same DNA construct—pSV2-CAT (Zoraqi and Spadafora, 1997; Magnano *et al.*, 1998; Maione *et al.*, 1998)—was used in all of these experiments, which implies the possibility of a sequence existing within the DNA construct that results in site-specific integration (Sperandio *et al.*, 1996). In contrast, a unique region "hot spot" may exist in the target cell genome at a specific time period of fertilization, the decondensation of the sperm nucleus, or the reconstruction of chromatin structure. The formation of female and male pronuclei may be highly accessible for foreign DNA integration (Zoraqi and Spadafora, 1997; Chan, 1999). The clustering of low and high gene transfer efficiency between experiments supports the aforementioned hypothesis (Maione *et al.*, 1998; Spadafora, 1998). Additional evidence against the integration event occurring before fertilization is the high mosaic rate in transgenic animals (Perez *et al.*, 1985; Lauria and Gandolfi, 1993; Habrova *et al.*, 1996; Chan *et al.*, 1999a). These results imply that integration occurs after the first DNA replication in the fertilized eggs rather than in the haploid genome of the sperm. A comparative study of integration sites in all transgenic species created by the sperm vector method could be very helpful in understanding the potential preferences and mechanisms of gene integration in sperm-mediated gene transfer.

V. Alternative Strategies

Successful sperm-mediated gene transfer to create transgenic animals requires three essential steps: DNA binding, internalization, and integration (Bachiller *et al.*, 1991; Lavitrano *et al.*, 1992; Lauria and Gandolfi, 1993; Zani *et al.*, 1995; Macha *et al.*, 1997; Magnano *et al.*, 199; Maione *et al.*, 1998; Spadafora, 1998). Several strategies have been devised to target these critical steps in order to improve gene transfer efficiency and to address the inconsistency of the technique.

4. Sperm-Mediated Gene Transfer

The use of a marker to evaluate DNA-binding efficiency on live sperm and to monitor the dynamics of sperm–DNA interaction will allow a better prediction of the rate of transgenesis (Chan *et al.*, 2000). The rhodamine-tagged DNA plasmid has been shown to be a powerful tool in monitoring and evaluating DNA-binding efficiency on live sperm. With the aid of this marking system, sperm carrying high or low amounts of DNA can be determined easily by the intensity of the rhodamine signal. Although binding efficiency can be evaluated before IVF, due to the fact that sperm has more intense rhodamine signal and tends to have lower motility, fertilization using highly labeled sperm may not be sufficient to improve the transgenic rate. We suggest the combination of DNA-bound sperm and intracytoplasmic sperm injection (ICSI) (TransgenICSI), together with the rhodamine-labeling system, as an alternative to the traditional gene transfer method: "pronuclear microinjection" (Chan *et al.*, 2000). Exogenous DNA transfer, mediated by ICSI with DNA-bound sperm, results in the production of rhesus embryos expressing the reporter gene: "green fluorescent protein" (GFP) (mean 34.6%; $N = 81$). Sperm bound with the rhodamine-tagged DNA plasmid can be selected under a fluorescent microscope. Selected sperm is used for ICSI, and the expression of GFP can be determined before embryo transfer. Although a live animal has been reported, transgenesis was not successful. Nonetheless, knowing the binding efficiency of DNA with live sperm provides a powerful tool in establishing criteria for sperm selection and improvement of transgenic efficiency. Damaging mouse sperm cell membrane by freeze-thaw, freeze-dry, or detergent treatment has demonstrated a significant improvement transgene expression in the resulting mouse embryos and has resulted in a consistent transgenic rate followed by ICSI (Perry *et al.*, 1999). The exposure of sperm inner cell membrane by the treatment just described may have revealed a region that is more susceptible for DNA binding, and the removal of inhibitory proteins on the sperm surface may favor the binding of DNA, perhaps facilitating the internalization of foreign DNA into the sperm nucleus. Indeed, when intact sperm is used, significantly lower numbers of embryos are expressing the transgene (Perry *et al.*, 1999). However, the rates of transgenesis in offspring using treated vs nontreated sperm have not been compared simultaneously. Because high variability in transgenic animal rates has been demonstrated, a control group using untreated sperm will strengthen the hypothesis (Maione *et al.*, 1998; Spadafora, 1998; Perry *et al.*, 1999). Although the binding of DNA on sperm has been well evaluated, a successful gene transfer may depend on the amount of DNA carried by the sperm prior to the fertilization process. The controversy between sperm motility and DNA-binding efficiency has been discussed previously (Bird *et al.*, 1992; Francolini *et al.*, 1993). The combination of ICSI and sperm vectors may create a bias during sperm selection and result in differences between *in vitro* and *in vivo* fertilization (Lauria and Gandolfi, 1993; Huguet and Esponda, 1998; Maione *et al.*, 1998). Indeed, the expression of exogenous DNA in embryos does not necessarily represent the actual integration event. Transient expression has been shown in embryos produced by the cytoplasmic injection of plasmid DNA with no transgenic animals produced (Brinster et al., 1985). A high GFP-expressing rate in a preimplantation embryo (>80%) compared with a lower transgenic animal rate (~21%) indicates a high transient expression rate in embryos following ICSI using DNA bound on damaged sperm (Perry et al., 1999). This result suggests that the majority of integration events do not occur prior to ICSI, and damage on the sperm cell surface may enhance the binding and internalization of DNA into the nucleus. Perry and colleagues (1999) success in creating transgenic mice has proven the idea of TransgenICSI. Although there is limited success in creating transgenic rhesus monkeys due to biological and ethical limitations, TransgenICSI holds promise as an alternative method for transgenic production.

Monitoring the internalization and integration of foreign DNA in sperm is not feasible without the analysis of the target cell genome or fixation of the sperm cell (Patil and Khoo, 1995; Zoraqi and Spadafora, 1997; Huguet and Esponda, 1998; Magnano et al., 1998; Farre et al., 1999; Chan et al., 2000). Although the amount of DNA that binds on the sperm cell surface may not be proportional to the amount of internalized DNA and the frequency of integration, the use of a DNA marker to select sperm with a defined amount of DNA will assist in the establishment of criteria for successful gene transfer by ICSI.

One of the key factors for the sperm vector system is the internalization of exogenous DNA into sperm nuclei. Various methods have been attempted in order to improve the internalization process. These include the use of liposomes (Bachiller et al., 1991; Nakanishi and Iritani, 1993), electroporation of spermatozoa (Gagne et al., 1991; Nakanishi and Iritani, 1993), and adenovirus-mediated gene transfer (Farre et al., 1999). Although a higher amount of foreign DNA was detected on sperm heads that resist DNase I treatment, a decrease in motility has been reported, and no further improvements in transgenic animal rates have been demonstrated using this technique (Bachiller et al., 1991; Nakanishi and Iritani, 1993; Farre et al., 1999). These results raise questions about the relationship between internalization and integration. Based on the random collision theory, DNA concentration plays an important role in the successful end joining of the injected DNA after pronuclear microinjection (Bishop and Smith, 1989; Bishop, 1996). Similarly, the binding of DNA on sperm cell membranes increases the local concentration of foreign DNA surrounding the future pronuclei, which facilitates intra- and intermolecular associations and the formation of concatemers (Bishop and Smith, 1989). Concatemers are believed to be the intermediate structures formed prior to the insertion of foreign DNA into chromosomes by an illegitimate route. In addition to DNA concentration, the integration process depends on the frequency of chromosomal breakage and the activation of the DNA repair mechanism (Bishop and Smith, 1989; Hamada et al., 1993; Bishop, 1996). Although DNA polymerase activity in mature sperm has been reported previously (Kramer and Krawetz, 1997), the repair mechanism ability has not yet been evaluated in sperm cells. The activity of DNA repair enzymes has been shown in amphibian eggs and embryos (Signoret and David, 1986). In addition to repair enzyme activity, irradiation on tissue culture cells increased gene transfer efficiency, which may be due to the increase of genomic DNA breakage or the induction of repair activity (Perez *et al.*, 1985). These reports suggest that successful DNA integration requires the aid of repair mechanisms. Thus, involvement of the oocyte in sperm-mediated gene transfer is suggested, although integration hot spots exist in the haploid genome of sperm (Kramer and Krawetz, 1997). Once again, this may suggest that the integration of foreign DNA occurs after fertilization.

The attempt to increase DNA-binding efficiency and the internalization rate seems to be not as effective as expected in creating transgenic animals. Although both enzyme activity and permissive sites within sperm nuclei have been demonstrated, compared to other somatic cells, the sperm cell is still considerably inert in most respects. Therefore, instead of relying on intrinsic activities within the sperm nucleus, an external aid may accelerate the integration rate, which is considered to directly affect the transgenic rate. The cotransfection of linearized plasmid DNA and restriction enzymes has resulted in a high transformation rate and specific integration sites in yeast, and this technique is referred to as restriction enzyme-mediated integration (REMI) (Schiestl and Petes, 1991; Kuspa and Loomis, 1992). Kuspa and Loomis (1992) suggested that the high integration rate is not simply the result of nonspecific double strand breaks, as restriction enzymes that create identical overhangs are required for the linearization of plasmid DNA. Because homologous ends are created in both the plasmid DNA and the target cell genome, the interruption by the linearized foreign DNA during the repair of target cell genome strand breaks may result in integration (Kuspa and Loomis, 1992). Although this enzymatic approach has been successful in yeast, the delivery of both plasmid DNA and restriction enzymes into sperm nuclei may be a challenge. Kroll and Amaya (1996) have adapted the REMI assay with essential modifications and have successfully created transgenic *Xenopus* embryos with high efficiency. *Xenopus* sperm nuclei were coincubated with restriction enzymes and followed by the addition of an interphase egg extract *in vitro*. The partially decondensed sperm nuclei were then transplanted into unfertilized oocytes (Kroll and Amaya, 1996). A high nonmosaic expression rate of the transgene was demonstrated. This indicates that the integration of plasmid DNA occurs before DNA replication and possibly during the process of nuclear decondensation (Kroll and Amaya, 1996). This strategy is similar to the method described by Perry and colleagues (1999). Both methods modify sperm and use an alternate strategy to deliver paternal chromatin into an oocyte to ensure the delivery of exogenous DNA. In fact, Perry and colleagues (1999) targeted damaged sperm cell membranes to enhance DNA binding, internalization, and perhaps integration. Compared with Perry and colleagues (1999), Kroll and Amaya (1996) targeted the integration of the foreign DNA in sperm nuclei by modifying the REMI assay. Integration was enhanced by partial decondensation of the sperm nucleus to expose the haploid genome and further creation of a unique overhang by specific restriction enzymes, which creates identical overhangs within the exogenous DNA (Kroll and Amaya, 1996). The creation of identical overhangs facilitates the integration of foreign DNA into the target cell genome as described by Kuspa and Loomis (1992). Although validation of the time of integration is not addressed, it is more likely to occur in the paternal nucleus because of the existing preference for integration and the high nonmosaic rate (Kroll and Amaya, 1996). Perhaps the integration process cannot be completed after the injection of sperm nuclei into oocytes, where functional repair machinery may exist.

VI. Perspectives

The concerns of using sperm as a carrier for pathogenic materials have long been raised (Brackett *et al.*, 1997; Levy *et al.*, 1980; Mulcahy and Pashco, 1984; Portis *et al.*, 1987; Kiessling *et al.*, 1989; Chan *et al.*, 1994; Baccetti *et al.*, 1998), even though the ability of sperm carrying exogenous DNA to create transgenic animals has not been clearly demonstrated (Castro *et al.*, 1990; Maione *et al.*, 1998; Spadafora, 1998). The binding of exogenous material to the sperm cell surface is a natural and nonselective event. Binding of exogenous DNA and viral particles has been demonstrated and studied extensively (Francolini *et al.*, 1993; Chan *et al.*, 1994, 1995; Zani *et al.*, 1995; Baccetti *et al.*, 1998). The removal of seminal plasma in the female genital tract before sperm reach the oocyte, and the fusion of sperm and egg membranes during fertilization, may not only serve as a selection process, but also for the removal of pathogenic materials in the plasma and on the sperm surface. However, success in using the sperm vector technique with either IVF or ICSI has raised the questions of potential consequences of assisted reproductive technologies. Both procedures bypass all of the natural defense mechanisms in the female genital tract, where potential pathogenic material may be removed during the fertilization process. Therefore, the delivery of unknown pathogens may occur. Nevertheless, today's breakthroughs in gamete gene transfer may lead to a new era of gamete gene therapy in which spermatozoa could be modified before fertilization to ensure that all the progeny cells carry functional copies of a gene. Although the advancement of such technology may lead us to new horizons of knowledge, the concerns of the consequences of this technology need to be critically addressed to ensure safe implementation.

Acknowledgments

We thank Drs. K. Y. Chong, Dr. Y. Agca, T. Dominko, and Mr. C. Payne for suggestions and reviewing of the manuscript. This work has been supported by NIH, NICHD, and NCRR.

4. Sperm-Mediated Gene Transfer

References

Atkinson, P. W., Hines, E. R., Beaton, S., Matthaei, K. I., Reed, K. C., and Bradley, H. P. (1991). Association of exogenous DNA with cattle and insect spermatozoa in vitro. *Mol. Reprod. Dev.* **29,** 1–5.

Baccetti, B., Benedetto, A., Collodel, G., di Caro, A., Garbuglia, A. R., and Piomboni, P. (1998). The debate on the presence of HIV-1 in human gametes. *J. Reprod. Immunol.* **41,** 41–67.

Bachiller, D., Schellander, K., Peli, J., and Ruther, U. (1991). Liposome-mediated DNA uptake by sperm cells. *Mol. Reprod. Dev.* **30,** 194–200.

Barone, J. G., de Lara, J., Cummings, K. B., and Ward, W. S. (1994). DNA organization in human sperm. *J. Androl.* **15,** 139–144.

Bird, J. M., Powell, R., Horan, R., Cannon, F., and Houghton, J. A. (1992). The binding of exogenous DNA fragments to bovine spermatozoa. *Anim. Biotechnol.* **3,** 181–200.

Bishop, J. O. (1996). Chromosomal insertion of foreign DNA. *Reprod. Nutr. Dev.* **36,** 607–618.

Bishop, J. O., and Smith, P. (1989). Mechanism of chromosomal integration of microinjected DNA. *Mol. Biol. Med.* **6,** 283–298.

Brackett, B. G., Baranska, W., Sawicki, W., and Koprowski, H. (1971). Uptake of heterologous genome by mammalian spermatozoa and its transfer to ova through fertilization. *Proc. Natl. Acad. Sci. USA* **68,** 353–357.

Brinster, R. L., Chen, H. Y., Trumbauer, M. E., Yagle, M. K., and Palmiter, R. D. (1985). Factors affecting the efficiency of introducing foreign DNA into mice by microinjecting eggs. *Proc. Natl. Acad. Sci. USA* **82,** 4438–4442.

Brinster, R. L., Sandgren, E. P., Behringer, R. R., and Palmiter, R. D. (1989). No simple solution to making transgenic mice. *Cell* **59,** 239–241.

Camaioni, A., Russo, M. A., Odorisio, T., Gandolfi, F., Fazio, V. M., and Siracusa, G. (1992). Uptake of exogenous DNA by mammalian spermatozoa: Specific localization of DNA on sperm head. *J. Reprod. Fertil.* **96,** 203–212.

Castro, F. O., Herna'ndez, O., Uliver, C., Solano, R., Milane's, C., Aguilar, A. Pe'rez, A., de Armas, R., Herrera, L., and De la Fuente, J. (1990). Introduction of foreign DNA into the spermatozoa of farm animals. *Theriogenology* **34,** 1099–1110.

Chan, A. W. S. (1999). Transgenic animals: Current and alternative strategies. *Cloning* **1,** 25–46.

Chan, A. W. S., Homan, E. J., Ballou, L. U., Burns, J. C., and Bremel, R. D. (1998). Transgenic cattle produced by reverse-transcribed gene transfer in oocytes. *Proc. Natl. Acad. Sci. USA* **95,** 14028–14033.

Chan, A. W. S., Kukoil, G., Skalka, A. M., and Bremel, R. D. (1999a). Timing of DNA integration, transgenic mosaicism, and pronuclear microinjection. *Mol. Reprod. Dev.* **52,** 406–413.

Chan, A. W. S., Luetjens, C. M., Dominko, T., Ramalho-Santos, J., Simerly, C. R., Hewitson, L., and Schatten, G. (2000). Foreign DNA transmission by intracytoplasmic sperm injection: Injection of sperm bound with exogenous DNA results in embryonic gene expression and live rhesus births. *Hum. Reprod.* **6,** 26–33.

Chan, P. J., Kalugdan, T., Su, B. C., Whitney, E. A., Perrott, W., Tredway, D. R., and King, A. (1995). Sperm as a noninvasive gene delivery system for preimplantation embryos. *Fertil. Steril.* **63,** 1121–1124.

Chan, P. J., Su, B. C., Kalugdan, T., Seraj, I. M., Tredway, D. R., and King, A. (1994). Human papillomavirus gene sequences in washed human sperm deoxyribonucleic acid. *Fertil. Steril.* **61,** 982–985.

Chevaillier, P., and Philippe, M. (1976). *In situ* detection of DNA-polymerase activity in the nuclei of mouse spermatozoa. *Chromosoma* **54,** 33–37.

Cody, C. W., Prasher, D. C., Westler, W. M., Prendergast, F. G., and Ward, W. W. (1993). Chemical structure of the hexapeptide chromophore of *Aequorea* green-fluorescent protein. *Biochemistry* **32,** 1212–1218.

Coffin, J. M. (1990). Molecular mechanism of nucleic acid integration. *J. Med. Virol.* **31,** 43–49.

Farre, L., Rigau, T., Mogas, T., García-Rocha, M., Canal, M., Gomez-Foix, A. M., and Rodríguez-Gil, J. E. (1999). Adenovirus-mediated introduction of DNA into pig sperm and offspring. *Mol. Reprod. Dev.* **53,** 149–158.

Francolini, M., Lavitrano, M., Lamia, C. L., French, D., Frati, L., Cotelli, F., and Spadafora, C. (1993). Evidence for nuclear internalization of exogenous DNA into mammalian sperm cells. *Mol. Reprod. Dev.* **34,** 133–139.

Gagne, M. B., Pothier, F., and Sirard, M. A. (1991). Electroporation of bovine spermatozoa to carry foreign DNA in oocytes. *Mol. Reprod. Dev.* **29,** 6–15.

Gandolfi, F., Lavitrano, M., Camaioni, A., Spadafora, C., Siracusa, G., and Lauria, A. (1989). The use of sperm-mediated gene transfer for the generation of transgenic pigs. *J. Reprod. Fertil.* Abstr **4,** 10.

Habrova, V., Takac, M., Navratil, J., Macha, J., Caskova, N., and Jonak, J. (1996). Association of Rous Sarcoma virus DNA with *X. laevis* spermatozoa and its transfer to ova through fertilization. *Mol. Reprod. Dev.* **44,** 332–342.

Hamada, T., Sasaki, H., Seki, R., and Sakaki, Y. (1993). Mechanism of chromosomal integration of transgenes in microinjected mouse eggs: Sequence analysis of genome-transgene and transgene-transgene junctions at two loci. *Gene* **128,** 197–202.

Hecht, N. B. (1974). A DNA polymerase isolated from bovine spermatozoa. *J. Reprod. Fertil.* **41,** 345–354.

Huguet, E., and Esponda, P. (1998). Foreign DNA introduced into the vas deferens is gained by mammalian spermatozoa. *Mol. Reprod. Dev.* **51,** 42–52.

Ivanova, M. M., Rosenkranz, A. A., Smirnova, O. A., Nikitin, V. A., Sobolev, A. S., Landa, V., Naroditsky, B. S., and Ernst, L. K. (1999). Receptor-mediated transport of foreign DNA into preimplantation mammalian embryos. *Mol. Reprod. Dev.* **54,** 112–120.

Khoo, H. W., Ang, L. H., Lim, H. B., and Wong, K. Y. (1992). Sperm cells as vectors for introducing foreign DNA into zebrafish. *Aquaculture* **107,** 1–19.

Kiessling, A. A., Crowell, R., and Fox, C. (1989). Epididymis is a principal site of retrovirus expression in the mouse. *Proc. Natl. Acad. Sci. USA* **86,** 5109–5113.

Kim, J. H., Jung-Ha, H. S., Lee, H. T., and Chung, K. S. (1997). Development of a positive method for male stem cell-mediated gene transfer in mouse and pig. *Mol. Reprod. Dev.* **46,** 1–12.

Kleene, K. C., Distel, R. J., and Hecht, N. B. (1984). Translational regulation and deadenylation of a protamine mRNA during spermatogenesis in the mouse. *Dev. Biol.* **105,** 71–79.

Kramer, J. A., and Krawetz, S. A. (1996). Nuclear matrix interactions within the sperm genome. *J. Biol. Chem.* **271,** 11619–11622.

Kramer, J. A., and Krawetz, S. A. (1997). RNA in spermatozoa: Implication for the alternative haploid genome. *Mol. Hum. Reprod.* **3,** 473–478.

Kroll, K. L., and Amaya, E. (1996). Transgenic *Xenopus* embryos from sperm nuclear transplantations reveal FGF signaling requirement during gastrulation. *Development* **122,** 3173–3183.

Kuspa, A., and Loomis, W. F. (1992). Tagging developmental genes in *Dictyostelium* by restriction enzyme-mediated integration of plasmid DNA. *Proc. Natl. Acad. Sci. USA* **89,** 8803–8807.

Lauria, A., and Gandolfi, F. (1993). Recent advances in sperm cell mediated gene transfer. *Mol. Reprod. Dev.* **36,** 255–257.

Lavitrano, M., Camaioni, A., Frati, V. M., Dolci, S., Farace, M. G., and Spadafora, C. (1989). Sperm cells as vectors for introducing foreign DNA into eggs: Genetic transformation of mice. *Cell* **57,** 717–723.

Lavitrano, M., French, D., Zanim, M., Frati, L., and Spadafora, C. (1992). The interaction between exogenous DNA and sperm cells. *Mol. Reprod. Dev.* **31,** 161–169.

Lavitrano, M., Maione, B., Forte, E., Francolini, M., Sperandio, S., Testi, R., and Spadafora, C. (1997). The interaction of sperm cells with exogenous DNA: A role of CD4 and major histocompatibility complex class II molecules. *Exp. Cell. Res.* **233,** 56–62.

Levy, J. A., Joyner, J., and Borenfreund, E. (1980). Mouse sperm can horizontally transmit type C viruses. *J. Gen. Virol.* **51,** 439–443.

4. Sperm-Mediated Gene Transfer

Luetjens, C. M., Payne, C., and Schatten, G. (1999). Non-random chromosome positioning in human sperm and sex chromosome anomalies following intracytoplasmic sperm injection. *Lancet* **353**, 1240.

Macha, J., Stursova, D., Takac, M., Habrova, V., and Jonak, J. (1997). Uptake of plasmid RSV DNA by frog and mouse spermatozoa. *Folia Biol. (Praha)* **43**, 123–127.

Magnano, A. R., Giordano, R., Moscufo, N., Baccetti, B., and Spadafora, C. (1998). Sperm/DNA interaction: Integration of foreign DNA sequences in the mouse sperm genome. *J. Reprod. Immunol.* **41**, 187–196.

Maione, B., Lavitrano, M., Spadafora, C., and Kiessling, A. A. (1998). Sperm-mediated gene transfer in mice. *Mol. Reprod. Dev.* **50**, 406–409.

Maione, B., Pittoggi, C., Achene, L., Loranzini, R., and Spadafora, C. (1997). Activation of endogenous nucleases in mature sperm cells upon interaction with exogenous DNA. *DNA Cell Biol.* **16**, 1087–1097.

Mori, T., Guo, M. W., Mori, E., Shindo, Y., Mori, N., Fukuda, I., and Mori, T. (1990). Expression of class II major histocompatibility complex antigen on sperm and its roles in fertilization. *Am. J. Reprod. Immunol.* **24**, 9–10.

Mulcahy, D., and Pashco, R. J. (1984). Adsorption to fish sperm of vertically transmitted fish viruses. *Science* **225**, 333–335.

Nakanishi, A., and Iritani, A. (1993). Gene transfer in the chicken by sperm-mediated methods. *Mol. Reprod. Dev.* **36**, 258–261.

Patil, J. G., and Khoo, H. W. (1995). Ultrastructural *in situ* hybridization and autoradiographic detection of foreign DNA in zebrafish spermatozoa. *Biochem. Mol. Biol. Int.* **35**, 965–969.

Perez, C. F., Botchan, M. R., and Tobias, C. A. (1985). DNA-mediated gene transfer efficiency is enhanced by ionizing and ultraviolet irradiation of rodent cells in vitro. *Radic. Res.* **104**, 200–213.

Perry, A. C. F., Wakayama, T., Kishikawa, H., Kasai, T., Okabe, M., Toygoda, Y., and Yanagimachi, R. (1999). Mammalian transgenesis by intracytoplasmic sperm injection. *Science* **284**, 1180–1183.

Portis, J. L., McAtee, F. J., and Hayes, S. F. (1987). Horizontal transmission of murine retroviruses. *J. Virol.* **61**, 1037–1044.

Rijkers, T., Peetz, A., and Ruther, U. (1994). Insertional mutagenesis in transgenic mice. *Transgen. Res.* **3**, 203–215.

Rottmann, O. J., Antes, R., Hoefer, P., and Maierhofer, G. (1991). Liposomes mediated gene transfer via spermatozoa into avian eggs cells. *J. Anim. Breed. Genet.* **109**, 64–70.

Schellander, K., Peli, J., Schmall, F., and Brem, G. (1995). Artificial insemination in cattle with DNA-treated sperm. *Anim. Biotech.* **6**, 41–50.

Schiestl, R. H., and Petes, T. D. (1991). Integration of DNA fragments by illegitimate recombination in *Saccharomyces cerevisiae*. *Proc. Natl. Acad. Sci. USA* **88**, 7585–7589.

Signoret, J., and David, J. C. (1986). DNA-ligase activity in axolotl early development: Evidence for a multilevel regulation of gene expression. *J. Embryol. Exp. Morphol.* **97**, 85–95.

Singh, G. B., Kramer, J. A., and Krawetz, S. A. (1997). Mathematical method to predict regions of chromatin attachment to the nuclear matrix. *Nucleic Acid Res.* **25**, 1419–1425.

Spadafora, C. (1998). Sperm cells and foreign DNA: A controversial relation. *BioEssays* **20**, 955–964.

Sperandio, S., Lulli, V., Bacci, M. L., Forni, M., Maione, B., Spadafora, C., and Lavitrano, M. (1996). Sperm-mediated DNA transfer in bovine and swine species. *Anim. Biotech.* **7**, 59–77.

Stewart, K. S., Kramer, J. A., Evans, M. I., and Krawetz, S. A. (1999). Temporal expression of the transgenic human protamine gene cluster. *Fertil. Steril.* **71**, 739–745.

Ward, W. S., Kimura, Y., and Yanagimachi, R. (1999). An intact sperm nuclear matrix may be necessary for the mouse paternal genome to participate in embryonic development. *Biol. Reprod.* **60**, 702–706.

Witkin, S. S., Korngold, G. C., and Bendich, A. (1975). Ribonuclease-sensitive DNA-synthesizing complex in human sperm heads and seminal fluid. *Proc. Natl. Acad. Sci. USA* **72**, 3295–3299.

Wu, G. M., Nose, K., Mori, E., and Mori, T. (1990). Binding of foreign DNA to mouse sperm mediated by its MHC class II structure. *Am. J. Reprod. Immunol.* **24,** 120–126.

Wykes, S. M., Nelson, J. E., Visscher, D. W,. Djakiew, D., and Krawetz, S. A. (1995). Coordinate expression of the PRM1, PRM2 and TNP2 multigene locus in human testis. *DNA Cell Biol.* **14,** 155–161.

Zalensky, A. O., Allen, M. J., Kobayashi, A., Zalenskaya, I. A., Balhorn, R., and Bradbury, E. M. (1995). Well-defined genome architecture in human sperm nucleus. *Chromosoma* **103,** 577–590.

Zani, M., Lavitrano, M. L., French, D., Lulli, V., Maione, B., Sperandio, S., and Spadafora, C. (1995). The mechanism of binding of exogenous DNA to sperm cells: Factors controlling the DNA uptake. *Exp. Cell Res.* **217,** 57–64.

Zoraqi, G., and Spadafora, C. (1997). Integration of foreign DNA sequences into mouse sperm genome. *DNA Cell Biol.* **16,** 291–300.

5

Gonocyte–Sertoli Cell Interactions during Development of the Neonatal Rodent Testis

Joanne M. Orth, William F. Jester, Ling-Hong Li, and Andrew L. Laslett
Temple University School of Medicine
Department of Anatomy and Cell Biology
Philadelphia, Pennsylvania 10140

I. The Perinatal Period of Testicular Development: Historical Perspective
   A. Germ Cells
   B. Sertoli Cells
II. The Sertoli Cell–Gonocyte Coculture Model
   A. Introduction
   B. Gonocyte Development in Coculture Mimics Development *in Vivo*
III. Cell–Cell Interactions in Neonatal Gonocyte Development
   A. The Role of *c-kit*
   B. Neural Cell Adhesion Molecule (NCAM)-Based Cell–Cell Adhesion in Neonates
   C. Developmental Pattern of NCAM Expression
   D. Regulation of NCAM Expression by T_3 and Its Impact on Gonocytes
   E. Other Potential Regulators of Gonocyte Function in Neonates
IV. Summary
   References

During neonatal testicular development in the rat, events critical for subsequent germ cell development occur that set the stage for fertility later in life. Some gonocytes resume mitotic activity and/or migrate to the surrounding basal lamina, and use of a carefully defined Sertoli cell–gonocyte coculture system indicates that these crucial events occur without added factors or hormones and are hence likely to depend on interaction with adjacent Sertoli cells. Coupling of the Kit receptor protein on gonocytes to stem cell factor from Sertoli cells is vital for successful migration by gonocytes, as antagonism of the former suppresses and addition of the latter stimulates gonocyte migration. During the neonatal period, intercellular adhesion is modified in a developmental manner such that neural cell adhesion molecule (NCAM) is the main adhesive molecule expressed and functioning at birth, with a progressive decline as development proceeds. This decline in NCAM expression is supported by the addition of exogenous 3,3′,5-triiodothyronine *in vitro*, and because this factor is recognized as supporting Sertoli cell differentiation, it seems likely that changing intercellular adhesion is a function of progressive development of Sertoli cells. Other avenues whereby maturing testicular cells influence each other doubtless exist, including secretion of growth factors and other peptides and developmentally important changes in the makeup of the extracellular matrix, which Sertoli cells and gonocytes contact. Continued investigation in these areas will be very valuable in enlarging our understanding of how neonatal testicular development provides the basis for successful spermatogenesis. © 2000 Academic Press.

I. The Perinatal Period of Testicular Development: Historical Perspective

A. Germ Cells

Although it is intuitively obvious that the proper development of germ cells in perinates is a prerequisite for normal sperm production in adults and that the "stem cells" responsible for lifelong male fertility arise from this population, our understanding of germ cell maturation in neonates and of the underlying regulatory mechanisms was poor until recent years. For example, although it has long been recognized that *fetal* germ cells are mitotically active during testicular morphogenesis, we knew little about mitotic activity in later fetal or neonatal germ cells beyond the observations made by Clermont and Perey (1957) in the late 1950s of large, apparent germ cell mitoses in 4-day-old rats. Similarly, *fetal* or primordial germ cells have long been known to migrate to the gonadal ridge where they become fully enclosed within the seminiferous cords between adjacent Sertoli cells. In addition, for many years it has been recognized that, subsequent to compartmentalization of the epithelium, their progeny, the spermatogonia, reside in the basal compartment contacting the basal lamina. However, beyond the early observation of neonatal germ cells or gonocytes with cellular extensions directed toward the tubule periphery (Huckins, 1963; Novi and Saba, 1968), for many years nothing was known about how the epithelium is remodeled and the germ cells repositioned postnatally. In our initial efforts to understand how these cells develop in neonates, we demonstrated that gonocytes in newborns resume motile behavior on day 4 after birth and migrate actively to reach the basal lamina and found that at least some of these cells, although not all, undergo division prior to their migration (McGuinness and Orth, 1992a,b), as shown in Fig. 1. Because of their obvious importance for ultimate fertility, our subsequent studies and those of others have focused on the regulation of these developmental events and on the relationship between gonocytes and Sertoli cells during this period.

B. Sertoli Cells

As is true for the germ cell population, perinatal development of Sertoli cells has been and to some extent still is somewhat poorly understood, despite the obvious importance of the developmental period when the size of this population is

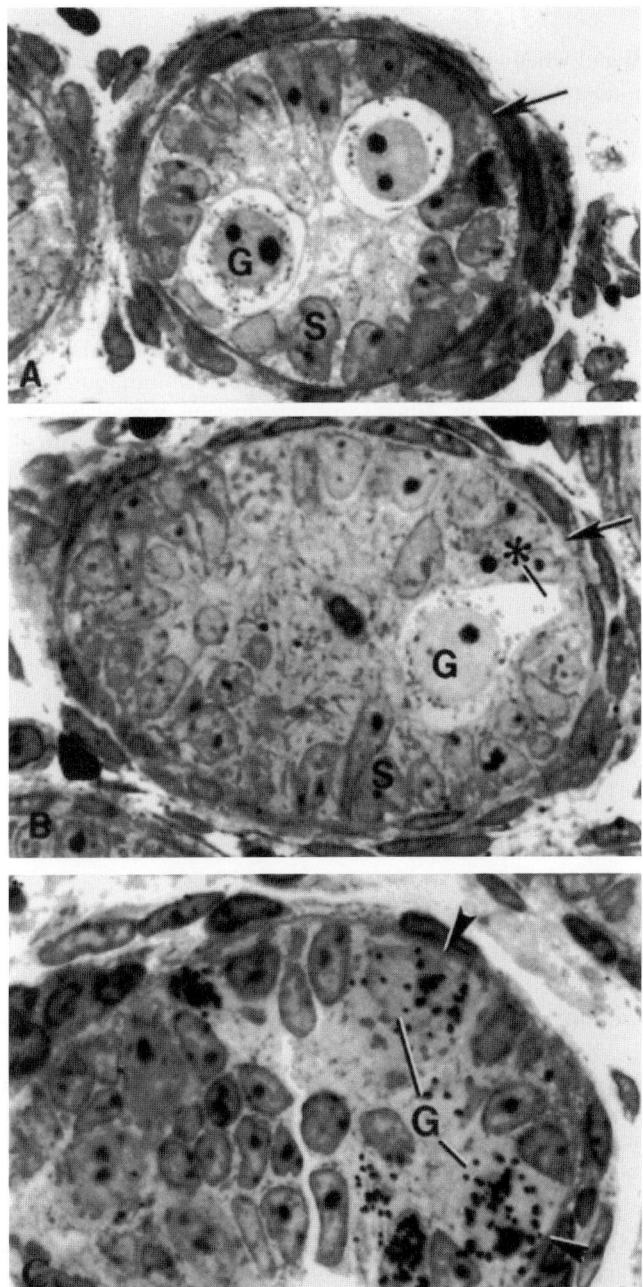

Fig. 1 Light micrographs of seminiferous cords from testes of rat pups on days 1 (A) and 4 (B and C) after birth. On the day of birth (A), all gonocytes (G) are round and separated from the basal lamina (arrow) by Sertoli cells (S). By day 4, some gonocytes have pseudopods (asterisk in B) directed toward the periphery whereas others have come into contact with the basal lamina (C). Reprinted with permission from McGuinness and Orth, *Anat. Rec.* **233**, 527–537 (1992). Reprinted by permission of Wiley-Liss, Inc., a subsidiary of John Wiley & Sons, Inc.

established and when terminal differentiation of these cells occurs. During the 1980s, our laboratory identified the final 2–3 days of gestation as the period during which the *rate* of Sertoli cell proliferation is maximal in rats (Orth, 1982). In addition, although the mitotic rate declines steadily postnatally, it became clear that the actual *size* of this population continues to increase during the first 7–10 days after birth. Subsequently, we demonstrated the essential role of follicle-stimulating hormone (FSH) in upregulating Sertoli cell proliferation and that β-endorphin from Leydig cells modifies responsiveness of these cells to FSH (Orth, 1986; Orth and Boehm, 1990a). Finally, work from the laboratory of Cooke (1994) with several *in vivo* and *in vitro* models identified thyroid hormone as a critical physiologic stimulator of Sertoli cell mitotic quiescence and accompanying differentiation. Thus, some information is available clarifying the regulation of Sertoli cell population size, a parameter shown by us and others to be critical in determining the quantitative output of sperm in the adult. In addition, we now recognize thyroid hormone as a major influence on the ability of Sertoli cells to proliferate and on their progression to terminal differentiation, a change in status sure to be of great importance for the adjacent germ cells. However, we know little of the molecular mechanisms that promote 3,3′,5-triiodothyronine (T_3)-induced Sertoli cell differentiation or of the likely other modifiers of this critical developmental event. Thus, there is a great need to explore and understand the functional maturation of both Sertoli and germ cells of neonates and to investigate how they interact during this time. The aim of this review, therefore, is to summarize the findings of recent studies focused on those interactions between Sertoli cells and gonocytes critical for these processes and on their regulation, as well as to highlight those important areas in need of further study.

II. The Sertoli Cell–Gonocyte Coculture Model

A. Introduction

Development of an *in vitro* model system in which to explore Sertoli cell–gonocyte interactions has made it possible to probe several aspects of neonatal development of these cells. First developed and described in 1990 (Orth and Boehm, 1990b), this approach involves isolation and pooling of seminiferous cords from newborn rats and then disruption of these cords via sequential enzymatic digestions, followed by harvesting of single cells from a dissociation buffer and their plating on an artificial basement membrane material, Matrigel. Characterization of these cultures indicated that approximately 90% of the cells are Sertoli cells, which rapidly form a confluent monolayer on the underlying substrate. The cultures also contain peritubular cells, constituting 5% or less of the total, and the remainder of the cells are gonocytes, easily recognizable by their characteristic morphology (see Figs. 2A and 2B). Initial characterization of these cocultures (Orth and Boehm, 1990b) revealed that cells maintained under these conditions retained most of the morphological characteristics seen *in vivo*. Moreover, use of glass bead loading and gap junction-permeant and nonpermeant fluorescent markers demonstrated that Sertoli cells and gonocytes are metabolically coupled in coculture (Fig. 3) and hence capable of physiological interaction. Thus, this *in vitro* approach affords an opportunity to study the functional relationship between these cell types under culture conditions that can be closely controlled and manipulated.

B. Gonocyte Development in Coculture Mimics Development *in Vivo*

During the first week of postnatal life, the gonocyte population in rats initiates a developmental program that will be crucial for the proper onset of spermatogenesis. In fetuses, gonocytes are mitotically quiescent from fetal day 17 onward (Orth, 1982). However, on postnatal day 3, approximately 10% of these cells enter S phase, and about the same number resume proliferation on each of the next 2 days (McGuinness and Orth, 1992). In addition, on postnatal day 4 some gonocytes complete their relocation and arrive at the basement membrane (Orth and McGuinness, 1992). Some of these migratory cells divide before relocating whereas others do not. Appropriate resumption of gonocyte mitosis and migration provides the foundation of further development of the gonocytes and their progeny, the type A spermatogonia. Thus, the first 3–5 postnatal days are of crucial importance in ensuring the successful onset of spermatogenesis and ultimate fertility of the testis in the adult.

Although initial, descriptive observations of gonocyte development were made *in vivo*, subsequent use of *in vitro* approaches has provided invaluable insights, enhancing our understanding of how the development of these cells is regulated. For example, when Sertoli cells and gonocytes are isolated and cocultured on the day of birth, both reinitiation of gonocyte division and their resumption of migratory behavior occur *spontaneously* and at time points roughly equivalent to the postnatal periods when each begins *in vivo* (McGuinness and Orth, 1992b). Thus, proliferation resumes on postnatal day 3 *in vivo* and, in cells cultured on the day of birth, on the third day in culture (Fig. 4). Similarly, migration to the basal lamina is achieved by the first gonocytes on postnatal day 4 and, in cocultured cells from newborns, cells with pseudopods first appear on the fourth day of culture (Fig. 2). Because few if any Leydig cells contaminate these cultures and because they are serum and hormone free, this strongly suggests that any factors influencing gonocyte mitotis and/or motility originate from the cells of the seminiferous tubules themselves or from products of these cells. Efforts from this laboratory and from several others have been concentrated on identifying and determining the cellular sources of those signals with a role in initiat-

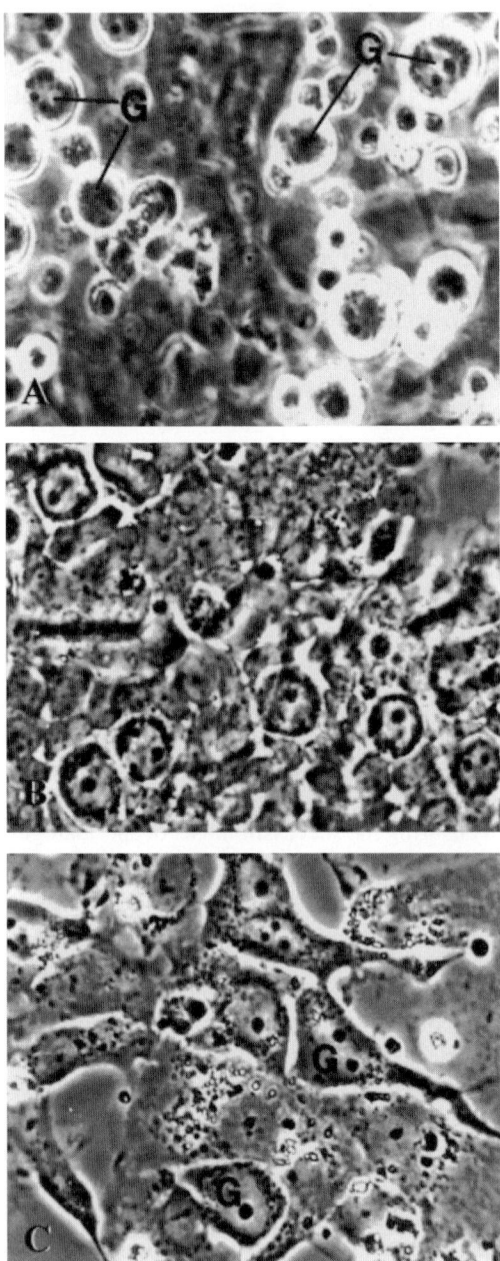

Fig. 2 Phase-contrast views of gonocytes (G) cocultured at birth and maintained for 1 (A), 2 (B), or 3 (C) days. In (A), all gonocytes are attached to Sertoli cells but still rounded up. By the second day (B), most have flattened and are at the same focal plane as the Sertoli cell monolayer and, by the third day in culture, many have developed pseudopods and display motile behavior (C). A and C reprinted with permission from Orth and McGuinness, *Endocrinology* **129,** 1119–1121 (1991). © The Endocrine Society.

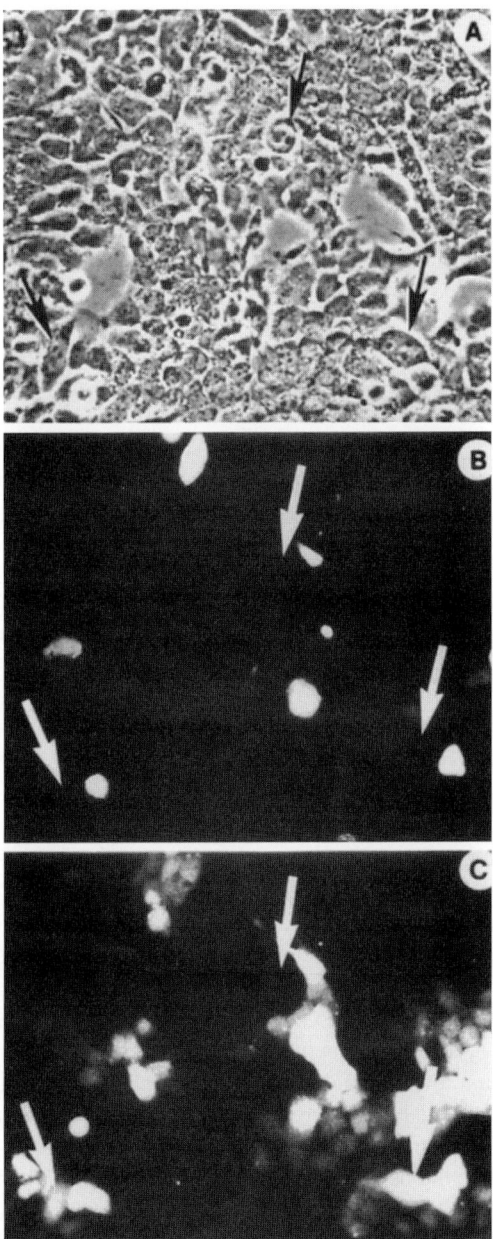

Fig. 3 Three views of the same area of coculture, showing phase contrast (A), rhodamine-dextran (B), and Lucifer yellow (C) fluorescence. Cells were bead loaded simultaneously with the two fluorescent probes, the former impermeant to gap junctions whereas the latter is permeable via these channels. The same three gonocytes are indicated by arrows in each of the three views. Each of these was not bead loaded directly as evidenced by the lack of signal in (B), but was adjacent to bead-loaded Sertoli cells. Note that passage of Lucifer yellow into two of the three gonocytes from the adjacent Sertoli cells can be seen in C. Reprinted with permission from Orth and Boehm, *Endocrinology* **127,** 2812–2820 (1990). © The Endocrine Society.

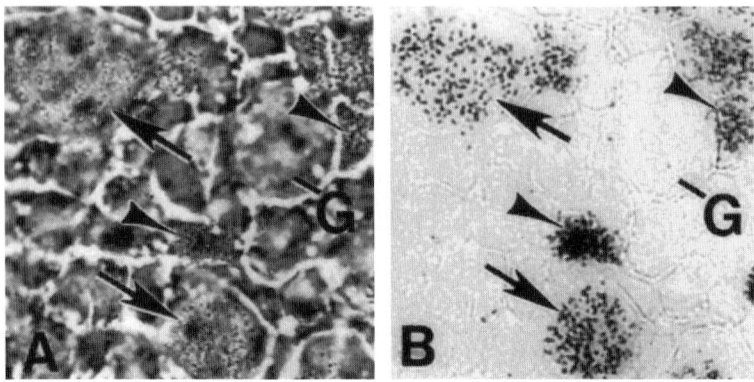

Fig. 4 Corresponding phase (A) and bright field (B) views of autoradiographs showing cells isolated at birth and cultured for 4 days. Arrows indicate labeled, proliferating gonocytes and arrowheads indicate mitotic Sertoli cells. Reprinted with permission from Orth and McGuinness, *Endocrinology* **129**, 1119–1121 (1991). © The Endocrine Society.

ing and regulating proliferation and migration of neonatal gonocytes. The remainder of this review focuses on the results of these endeavors and will identify those areas where critical information is still lacking.

III. Cell–Cell Interactions in Neonatal Gonocyte Development

Considering the close physical and functional relationship that is now recognized to exist between Sertoli cell and germ cells, it is not surprising that the most profound influences on gonocyte development in neonates appear to originate from the nearby Sertoli cell. In addition, evidence that these cells adhere avidly to each other *in vivo* as well as *in vitro* stems from our earlier observations in testes exposed to hypertonic buffer prior to fixation. In this tissue, we detected the presence of multiple sites of attachment between these two cell types that were highly resistant to separation, even though substantial shrinkage spaces appeared between most cells following the hypertonic treatment (Fig. 5). Moreover, besides providing for cell–cell attachment, these and other intercellular contacts have an obvious potential to regulate the behavior of one or both cell types. Thus far, we have explored the role of two surface factors, Kit protein and neural cell adhesion molecule (NCAM), during development of germ cells of neonates, as detailed later.

A. The Role of *c-kit*

Several naturally occurring mutations in the genes encoding the tyrosine kinase receptor protein Kit or its ligand stem cell factor (SCF) produce a virtual absence

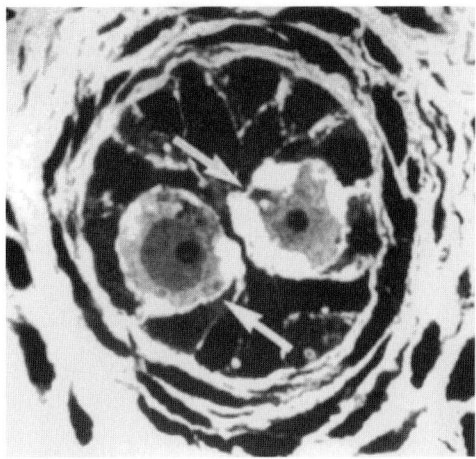

Fig. 5 A view of a seminiferous cord from a 3-day-old rat pup perfused *in vivo* with hypertonic fixative to induce tissue shrinkage and intercellular separation. Arrows indicate examples of areas of tight, shrinkage-resistant fusion between each of two gonocytes and adjacent Sertoli cells.

of germ cells in the developing testes, leading to infertility (for review, see Loveland and Schlatt, 1997). Early studies on these genes suggested that primordial germ cells do not colonize the gonad in Kit-deficient mutants because they either fail to migrate (Mintz and Russell, 1957) or to proliferate (McCoshen and McCallion, 1975). Subsequent data pointed to a definitive role for SCF as a "survival factor" for primordial germ cells rather than a direct stimulator of their proliferation (Godin *et al.*, 1991; Dolci *et al.*, 1991). Moreover, in later studies for which antisera directed against the Kit protein were injected into postnatal mice, depletion of spermatogonia and spermatocytes was observed in adults (Yoshinaga *et al.*, 1991) and this was attributed to enhanced apoptosis of these cells following the neutralization of Kit (Packer *et al.*, 1995). Thus, the concept of the Kit receptor and its ligand as part of a survival system for some or all undifferentiated germ cells has become widely held.

In light of these and other observations, we asked whether Kit–SCF interaction might have a role in Sertoli cell-mediated regulation of gonocyte development in *neonatal* rats. After first determining by Northern analysis that the *c-kit* gene is expressed in neonatal tests between postnatal days 1 and 5, we used *in situ* hybridization to localize its message (Orth *et al.*, 1996). We found that most if not all gonocytes express *c-kit* at birth and that even a stronger signal is detectable in virtually all of these cells by day 5, a time when both mitosis and migration have been reinitiated in the population (Fig. 6A). In addition, we confirmed by Western analysis and immunolocalization that the protein is present in these cells and that it is found primarily at their surfaces (Fig. 6B). Thus, the Kit receptor protein is produced by neonatal gonocytes during the time when their postnatal migration is initiated and their division resumes.

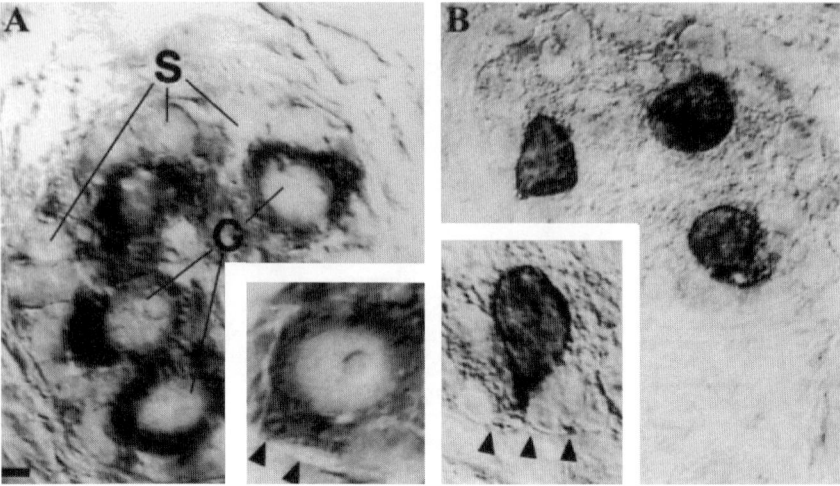

Fig. 6 Paraffin-embedded section subject to *in situ* hybridization (A) and frozen section subjected to immunolocalization (B) to visualize mRNA for *c-kit* and the Kit receptor protein, respectively, on postnatal day 5. In each panel, a gonocyte either in contact with the basal lamina (A) or apparently migrating toward the basal lamina (B) can be seen in the inserted view. Most if not all gonocytes express this gene on day 5 and contain abundant Kit protein, with a concentration of protein at the cell surface. Reprinted with permission from Orth *et al.*, *Mol. Reprod. Dev.* **45**, 123–131 (1996). Reprinted by permission of Wiley-Liss, Inc., a subsidiary of John Wiley & Sons, Inc.

In subsequent analyses, we used the Sertoli cell–gonocyte coculture system to probe the potential role of Kit and its ligand, SCF, in development of these neonatal germ cells *in vitro* (Orth *et al.*, 1996, 1997). When *in situ* hybridization was applied to cells prepared on the day of birth and analyzed daily up to the fifth day *in vitro*, we detected some gonocytes positive for *c-kit* mRNA on each day, with an apparent upregulation in the numbers of cells strongly expressing the gene with increasing time in culture. This qualitative observation was confirmed quantitatively (not shown here), and when we asked whether *c-kit* expression is related to the presence or absence of pseudopods on the cells, we found an absolute correlation between migratory activity and *c-kit*-positive gonocytes on days 3 through 5 *in vitro* (Fig. 7), corresponding to the time of initiation and maintenance of motility among these cells. Because these findings suggested that the expression of *c-kit* might be required for migration to occur, we asked directly whether blockade of the receptor protein would prevent gonocytes from forming pseudopods by incubating cocultured cells with specific antisera against Kit. The results of quantitative analyses of cultures with and without exposure to this antisera indicated that, on days 4 through 6 after plating cells from newborns, approximately 30–40% fewer gonocytes with pseudopods were seen in antisera-

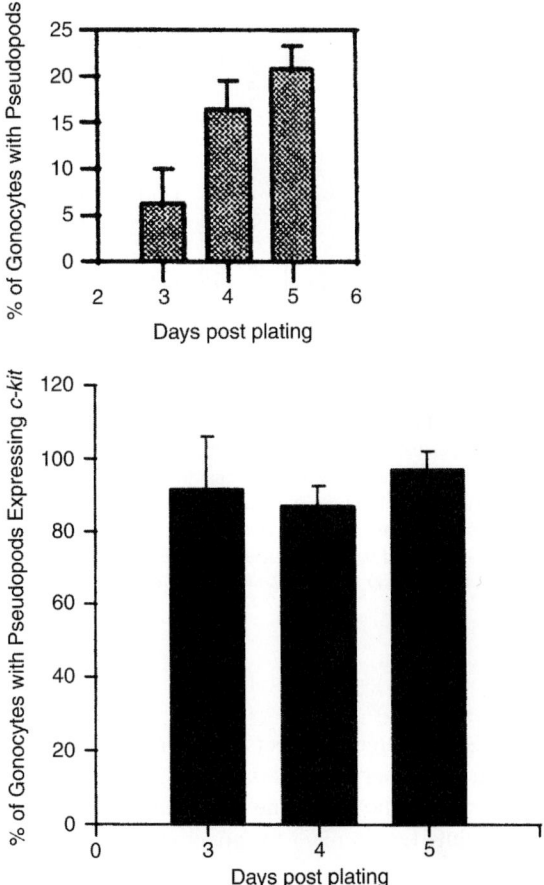

Fig. 7 The percentage of migratory gonocytes isolated on day 1 that express *c-kit* in cocultures fixed and subjected to *in situ* hybridization on days 3, 4, and 5 *in vitro*. Essentially all migratory cells were positive for *c-kit* mRNA. The inserted graph shows the percentage of cells that become migratory on days 3 through 5; note that no cells develop pseudopods before the third day of culture. Reprinted with permission from Orth *et al.*, *Biol. Reprod.* **57,** 676–683.

treated chambers compared to controls (Fig. 8, left). This decrease in migratory ability occurred without any change in the numbers of gonocytes in the treated cultures, strongly suggesting that the blockade of Kit action has no immediate and direct effect on gonocyte proliferation and/or survival, at least *in vitro* (not shown). Conversely, we also asked whether the exposure of cultured gonocytes to exogenously added SCF would *enhance* their migration *in vitro*. We found that cells exposed to SCF displayed a transient but significant increase in gonocytes

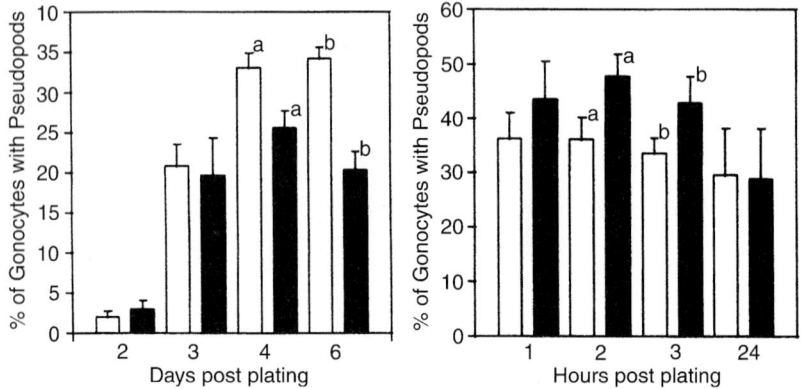

Fig. 8 (Left) The percentage of gonocytes with pseudopods in cultures exposed to Kit antiserum (solid bars) or control peptide (open bars) from the time of plating on postnatal day 1 through day 6 *in vitro*. Analysis of cell numbers in these dishes indicated that there were no differences in numbers of gonocytes between treated and control cultures (not shown). (Right) The percentage of gonocytes with pseudopods in cocultures prepared on day 5 after birth, when migration is ongoing *in vitro*, and exposed to either stem cell factor (solid bars) or vehicle (open bars) from 1 through 24 hr postplating. Reprinted with permission from Orth *et al.*, *Biol. Reprod.* **57,** 676–683 (1997).

with pseudopods in SCF-treated chambers compared to controls (Fig. 8, right). Thus, taken together, our observations on cocultured gonocytes provide strong evidence that expression of the *c-kit* gene and function of the encoded receptor Kit protein is an absolute requirement for gonocytes to reestablish their migratory activity after birth.

Our finding that gonocyte migration is suppressed *in vitro* following Kit suppression without any abnormal loss of cells from the cultures argues against a *direct* role for the Kit protein in avoidance of apoptosis in these germ cells. This suggestion is not inconsistent with the conclusion of others (e.g., Yoshinaga *et al.*, 1991; Packer *et al.*, 1995) that Kit is a "survival factor" as it has been recognized for many years that gonocytes that fail to migrate to the basal lamina in the early postnatal period do indeed die (Clermont and Perey, 1957). Thus, failure to reestablish motility in gonocytes when they are prevented from expressing *c-kit in vivo* may lead to their demise at a somewhat later time, via mechanisms presumably leading at that time to the onset of apoptosis and/or necrosis. This suggestion is supported by the fact that morphologically normal, centrally located gonocytes, which are presumably destined for eventual death, can be detected in seminiferous cords of developing testes long after successful migration has been completed by other members of this population (J. M. Orth, unpublished observations). A full understanding of the precise mechanisms underlying Kit action in developing gonocytes awaits a more detailed analysis of the molecular cascade of events triggered by Kit activation and of its impact *in vivo*.

B. Neural Cell Adhesion Molecule (NCAM)-Based Cell–Cell Adhesion in Neonates

Based on initial observations made in cocultures of neonatal Sertoli cells and gonocytes (Orth and Jester, 1995), it appears that the predominant adhesive mechanism attaching the former both to each other and to adjacent gonocytes is noncadherin in nature and involves NCAM. This factor, originally identified in neural tissue (Thiery, 1982) and subsequently in several nonneural tissues (Covault and Sanes, 1986; Mayerhofer *et al.*, 1991; Nouwen *et al.*, 1993), produces avid intercellular adhesion even at reduced temperatures or in the absence of significant amounts of Ca^{2+}. NCAM is posttranscriptionally modified into three isoforms, which display differences in the size of the cytoplasmic domain of the molecule. Two of these isoforms span the plasma membrane and thus have the potential to interact with the underlying cytoskeleton. NCAM also may be modified by the addition of polysialic acid to its N-terminal, extracellular domain, providing a further mechanism whereby the molecule can modify cell–cell spacing and hence affect other forms of intercellular interaction (for review, see Rutishauser, 1991). These attributes make NCAM a potentially important molecule in systems where flexibility in cell–cell interactions may be critical during important developmental stages.

In our studies, we characterized the molecular nature of neonatal testicular NCAM and found (1) that it is present exclusively as the 140-kDa isoform and (2) that it lacks polysialic acid on its N terminus (Li *et al.*, 1998). This is consistent with the presumed tight adhesion required between locomoting cells and their substrate and would also allow for NCAM–cytoskeletal linkages that could have a role in developmentally regulated gene expression in one or both cell types. In addition, immunolocalization in sections of testes on days 1 through 5 after birth indicated the presence of NCAM at most, if not all, Sertoli–Sertoli and Sertoli–gonocyte interfaces, with a similar localization also obvious in cocultured cells (Fig. 9). That NCAM functions in maintaining adhesion between gonocytes and Sertoli cells, at least *in vitro*, was demonstrated conclusively by the use of NCAM antiserum *in vitro* (Orth and Jester, 1995). In that study, we found (1) that the addition of anti-NCAM to cells from newborns shortly after plating interfered with the ability of gonocytes to form stable attachments to Sertoli cells and (2) that the addition of anti-NCAM to cocultures 4 days after plating, when attachment was well established, resulted in an obvious rounding up of gonocytes by 24 hr later (Fig. 10). In this way, we confirmed NCAM is phys-

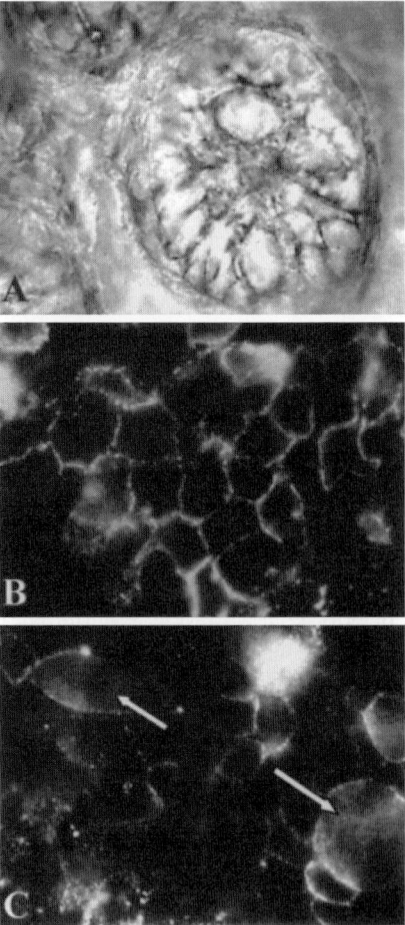

Fig. 9 Immunolocalization of NCAM in frozen sections on postnatal day 1 (A) or in cocultures prepared on postnatal day 5 (B and C). The plane of focus is at the level of the Sertoli cell monolayer in B and at the level of gonocytes in C, where arrows indicate both a round gonocyte and a gonocyte with a pseudopod. B and C reprinted with permission from Li et al., J. Androl. **19**, 365–373 (1998).

iologically important in maintaining adhesion between neonatal gonocytes and Sertoli cells, at least *in vitro*, and that interfering specifically with its function is sufficient to induce detachment of these cells.

C. Developmental Pattern of NCAM Expression

Examination of the developmental pattern of testicular NCAM expression *in vivo* (Li *et al.*, 1998) indicated that it is produced at highest levels during the first week

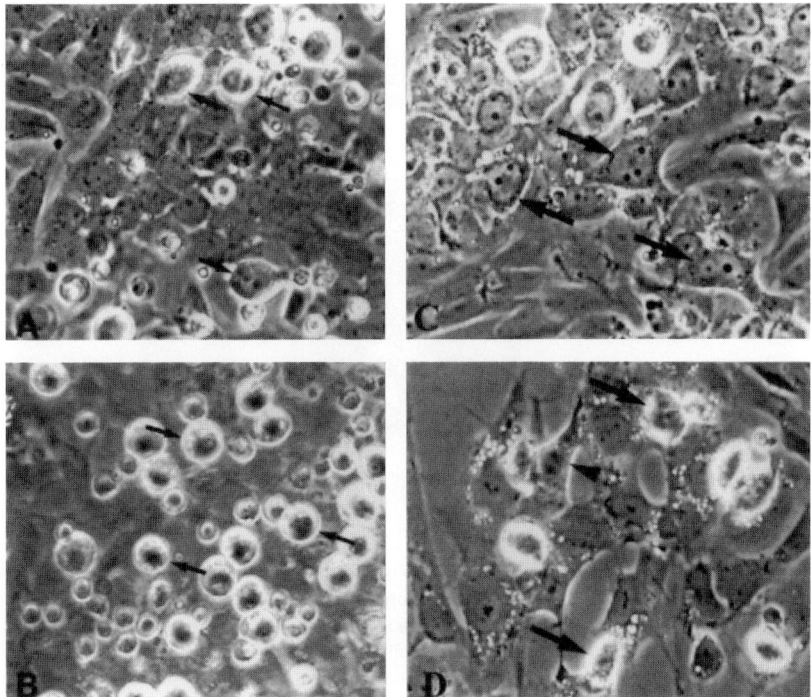

Fig. 10 The effect of treating cocultures with NCAM antiserum. (A and B) A culture shortly after plating, when gonocytes (arrows) are beginning to attach to underlying Sertoli cells, and the same culture approximately 24 hr later, after the addition of anti-NCAM at 1 hr postplating. (C and D) An untreated culture 4 days after plating (C) is compared to a view of the same culture after exposure to anti-NCAM for the subsequent day *in vitro*, with gonocytes indicated by arrows. Reprinted with permission from Orth *et al.*, *Biol. Reprod.* **57**, 676–683 (1997).

of postnatal life, with a diminished presence by 15 days and an absence in adults (Fig. 11). In contrast, we found that the pattern of expression *in vitro* is quite different, with a progressive *increase* in NCAM expression in cocultures maintained for up to 15 days (Li *et al.*, 1998). This observation suggests that the decrease in NCAM expression detected *in vitro* is influenced by non-Sertoli, perhaps extratesticular factors, a possibility we have been exploring in recent investigations.

A number of extratesticular factors, such as the gonadotropin hormones, have long been recognized as important during normal testicular development. However, a substantial body of evidence has accumulated in the last few years implicating thyroid hormone (T_3) as particularly important in influences on the development of the Sertoli cell population. For example, Sertoli cell differentiation is delayed in hypothyroid rats and is accelerated in the presence of hyperthy-

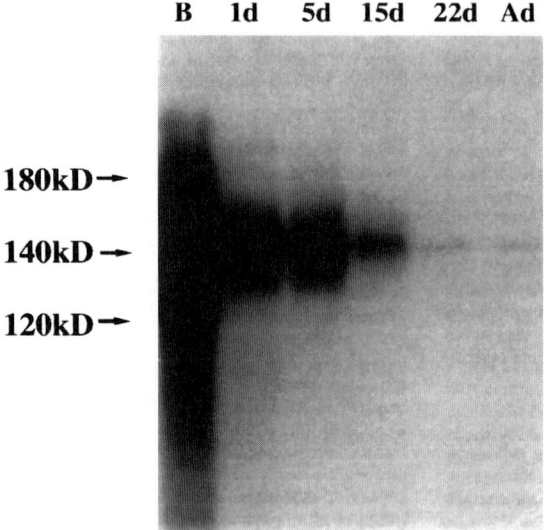

Fig. 11 Western blot of testicular protein samples obtained on days 1 through 22 postnatal and from adults, probed with an antibody recognizing all isoforms of NCAM. Protein from brain (B) was included as a positive control. Note that the expression of NCAM is highest during the first postnatal week and declines dramatically by day 15 and that all testicular NCAM is the 140-kDa isoform devoid of polysialic acid (PSA), in contrast to that from brain where the PSA moiety causes "smearing" of the immuno signal. Reprinted with permission from Li *et al.*, *J. Androl.* **19,** 365–373 (1998).

roidism (Van Haaster *et al.*, 1993; Jannini *et al.*, 1993; Cooke *et al.*, 1994; Simorangkir *et al.*, 1995, 1997), findings attributed to an effect of T_3 on FSH responsiveness and onset of differentiation among these cells (Cooke *et al.*, 1994). In addition, T_3-induced changes in the Sertoli cell population are accompanied by changes in numbers of gonocytes during development (Simorangkir *et al.*, 1997) as well as in numbers of germ cells in adults (Jannini *et al.*, 1993; Simorangkir *et al.*, 1995). Because germ cells appear to lack receptors for T_3, these changes in germ cell numbers are thought to reflect alterations in the size of the Sertoli cell population, a result in keeping with evidence indicating that the number of germ cells in the testis is directly linked to the size of the Sertoli cell population (Orth *et al.*, 1988). Work from our laboratory has provided additional data on a potential mechanism whereby T_3 impacts on germ cell development in neonatal testes, as follows.

D. Regulation of NCAM Expression by T_3 and Its Impact on Gonocytes

NCAM expression *in vivo* is highest early in postnatal life, with a progressive decline thereafter. Moreover, others have documented the presence of cadherins, notably P-cadherin, in developing rodent testes (Cyr et al., 1992; Byers et al., 1994; Lin and DePhilip, 1996). This information, taken together with data described earlier implicating T_3 as critical for Sertoli cell differentiation, led us to ask whether T_3 might have a role in the differential expression of adhesion factors by Sertoli and/or germ cells during neonatal development. To this end, we first exposed long-term Sertoli cell–gonocyte cocultures, in which we had earlier observed a *sustained* production of NCAM, to either T_3 or vehicle and noted that the presence of the hormone resulted in a virtual loss of detectable NCAM by 2 weeks of culture (Fig. 12). In addition, when T_3 was included for 4 days in cultures established on day 1, immunolocalization of NCAM indicated an obvious loss of signal from cell–cell boundaries (Fig. 13). Thus, T_3 suppresses NCAM expression in Sertoli cells, at least under the tested conditions *in vitro* (Laslett et al., 2000), suggesting that specific downregulation of NCAM by T_3 is a component of Sertoli cell differentiation during postnatal development. Moreover, this change in NCAM expression induced by T_3 also has a dramatic effect on gonocytes, at least *in vitro*. In our study, we detected an obvious detachment of germ cells from Sertoli cells in the treated chambers, and subsequent quantitative analysis of cells indicated that this detachment occurs in a dose- and time-dependent manner without any concomitant change in numbers of Sertoli cells in the underlying monolayer (Laslett et al., 2000). In addition, very low levels of T_3 appear to downregulate NCAM prior to any detachment of gonocytes, implying that NCAM downregulation *precedes* loss of gonocytes. Finally and perhaps most interesting, even in the virtual absence of NCAM

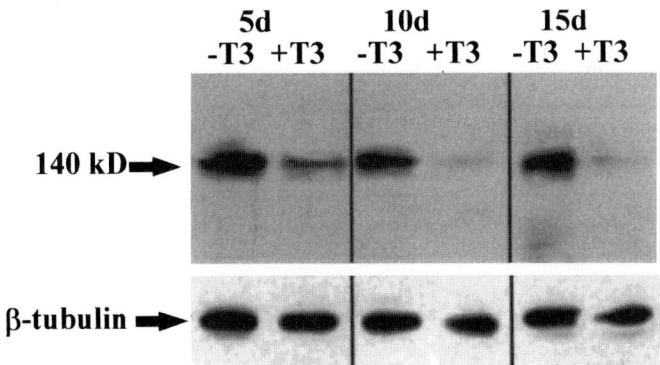

Fig. 12 Western analysis of NCAM in protein samples isolated from cocultures after 5–15 days of incubation with or without 100 nM T3. Protein was equally loaded on a 7.5% gel, subjected to PAGE, transferred, and immunoblotted with anti-NCAM. Equal loading of protein across lanes was verified by reprobing the blot for β-tubulin. A clear-cut downregulation of NCAM is induced by T_3, without any notable loss of Sertoli cells (not shown). Reprinted with permission from Laslett et al., *Endocrinology*, 2000.

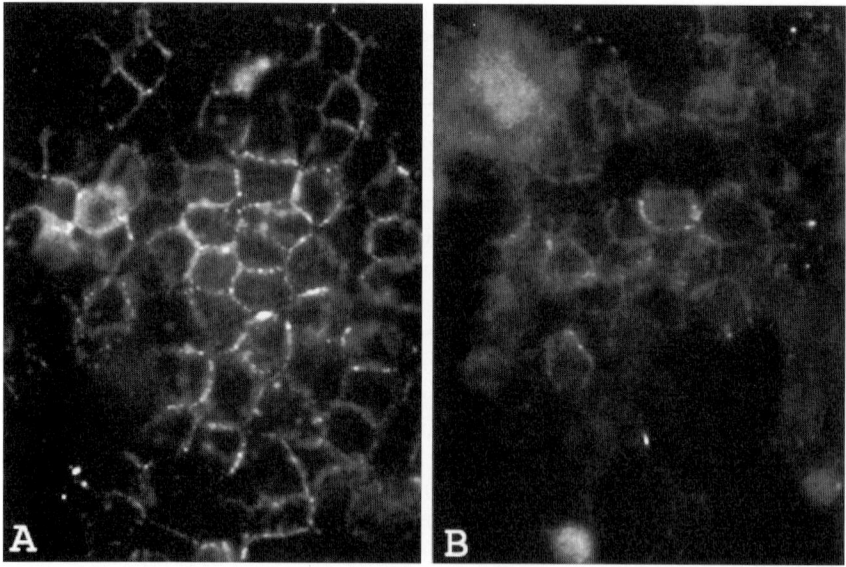

Fig. 13 Immunofluorescent localization of NCAM in cocultures maintained for 48 hr in the absence (A) or presence (B) of 0.1 nM T3. Note that the obvious and bright fluorescence at cell–cell boundaries is largely abolished by the presence of the hormone. Reprinted with permission from Laslett *et al.*, *Endocrinology*, 2000. © The Endocrine Society.

after extended T_3 treatment, approximately 20% of cocultured gonocytes retain their adhesion to the underlying Sertoli monolayer, suggesting that these adherent cells may represent a subset of gonocytes with the potential to express alternative adhesion factors and hence to survive. Considering the recognized role of Sertoli cells in determining the numbers of spermatogonia later in development, one might speculate that interaction between Sertoli cells and gonocytes in neonates is crucial for identifying the cohort of gonocytes destined to survive and that this may be regulated, at least in part, by T_3-induced changes in cellular adhesion. This interesting possibility will be well worth pursuing in subsequent studies aimed at understanding in detail the nature of these cell–cell interactions.

E. Other Potential Regulators of Gonocyte Functions in Neonates

1. Growth Factors

Although the nature of factors responsible for the *initiation* of either gonocyte migration or mitotic activity in neonates remains undetermined, some evidence has accumulated to support a role for several growth factors in modifying gonocyte proliferation subsequent to its onset. For example, the addition of exogenous leukemia inhibitory factor to Sertoli cell–gonocyte cultures resulted after several days in enhanced gonocyte mitotic activity (DeMiguel et al., 1996). In addition, quantitatively measurable increases in the uptake of Bromodeoxyuridine (BuDR) by these cells were detected following their exposure to either platelet-derived growth factor or estradiol (Li et al., 1997). At least for estradiol, it is clear that the developing Sertoli cells are a rich source of endogenous steroid and hence have the potential, via this route, of influencing gonocyte function. Moreover, beyond measurable effects on proliferation itself, these and several other factors, including fibroblast growth factor-2 (Van Dissel-Emiliani et al., 1996) and ciliary neurotropic factor (De Miguel et al., 1996), also seem able to upregulate the ability of gonocytes to avoid apoptosis and survive. Although the precise mechanism whereby these factors act *in vivo* has not yet been described, it seems likely that the Sertoli cell may be the source for at least some of them, providing an additional paracrine route for the control of germ cell development in neonates. Additional information will likely be forthcoming as investigations continue into these important questions.

2. Extracellular Matrix

Peritubular myoid cells are technically not components of the seminiferous epithelium and are not in contact with either Sertoli cells or gonocytes. However, they, along with Sertoli cells, contribute to the extracellular matrix of the basal lamina and thus have the potential to influence the function of either or both of these cells. Moreover, the role of the extracellular matrix in influencing testicular cell behavior is well recognized, especially during development (Hadley et al., 1990). Thus, the potential for changes in the makeup of the seminiferous basal lamina to influence the development of those cells that contact it in neonates is obvious. In an earlier study from our laboratory, we observed that the migratory behavior of gonocytes was highest when cells were cocultured on pure laminin compared to bare plastic, Matrigel, or fibronectin. Conversely, plating on Matrigel rather than laminin resulted in enhanced levels of gonocyte proliferation (Orth and McGuinness, 1991). In subsequent studies from other laboratories, the developmental regulation of other matrix components, notably the five type IV collagen α chains and the fibulins, has been explored with interesting results. For example, α 3 chains are first deposited into the basal lamina on postnatal day 5 (Enders et al., 1995), corresponding closely to the initial arrival of migratory gonocytes at that location. However, fibulin-1 is lost from the basal lamina after day 5 whereas fibulin-2 is localized in a nonrandom, segmental pattern by days 10–15 (Loveland et al., 1998). While only suggestive at this point in time, these and other observations on the basal lamina and its components in neonates raise the interesting possibility that differential indirect interaction between gonocytes and peritubular cells via the matrix may be important in influencing the subsequent survival and development of germ cells, a possibility well worth exploring in future studies.

IV. Summary

The importance of the neonatal period of testicular development for ensuring fertility at adulthood cannot be overestimated, as it is during this time that the initial cohort of stem cells responsible for producing the first generation of type A spermatogonia is established. Findings thus far have implicated the Sertoli cell as the source of at least some of the signals responsible for triggering postnatal development of these cells. In particular, SCF of Sertoli cell origin interacts with the Kit receptor on gonocytes to ensure successful migration of these cells, at least *in vitro*. The observation of universal cell death among those gonocytes that do not successfully migrate to the basal lamina *in vivo* identifies SCF–Kit interaction as likely to be critical for the proper development of postnatal gonocytes.

Adhesion between Sertoli cells and gonocytes in the postnatal period is also developmentally regulated in that NCAM expression is progressively downregulated during the first postnatal week. This downregulation seems to be supported by T_3 through its interaction with Sertoli cells and presumably involves differential signaling between these cells and gonocytes. This change in adhesion with progressing development is doubtless critical in identifying those germ cells destined to survive and to maintain an association with terminally differentiating Sertoli cells and is an area of research likely to yield exciting new data in future studies.

Finally, both secreted factors from Sertoli cells, nearby peritubular cells, or even from the germ cells themselves may be critical for appropriate development during the neonatal period. While most if not all studies on growth factors and matrix molecules thus far have been conducted *in vitro*, the findings provide important clues to the identity of important factors *in vivo*. Thus, extension of these studies to the living animal would be very valuable, providing meaningful data about other intercellular mechanisms that may not involve direct cell–cell contact but that may be crucial for testicular development in neonates.

References

Byers, S. W., Sujarit, S., Jegou, B., Butz, S., Hoschutzky, H., Herrenknecht, K., MacCalman, C., and Blaschuk, O. W. (1994). Cadherins and cadherin-associated molecules in the developing and maturing rat testis. *Endocrinology* **134,** 630–639.
Clermont, Y., and Perey, B. (1957). Quantitative study of the cell population of the seminiferous tubules in immature rats. *Am. J. Anat.* **100,** 241–260.

5. Sertoli Cells and Gonocytes in Neonates

Cooke, P. S., Zhao, Y. D., and Bunick, D. (1994). Triiodothyronine inhibits proliferation and stimulates differentiation of cultured neonatal Sertoli cells: Possible mechanism for increased adult testis weight and sperm production induced by neonatal goitrogen treatment. *Biol. Reprod.* **51**, 1000–1005.

Covault, J., and Sanes, J. R. (1986). Distribution of N-CAM in synaptic and extrasynaptic portions of developing and adult skeletal muscle. *J. Cell. Biol.* **102**, 716–730.

Cyr, D. G., Blaschuk, O. W., and Robaire, B. (1992). Identification and developmental regulation of cadherin messenger ribonucleic acids in the rat testis. *Endocrinology* **131**, 139–145.

DeMiguel, M. P., De Boer-Brouwer, M., Paniagua, R., Van Den Hurk, R., De Rooij, D. G., and Van Dissel-Emiliani, F. M. (1996). Leukemia inhibitory factor and ciliary neurotropic factor promote the survival of Sertoli cells and gonocytes in coculture system. *Endocrinology* **137**, 1885–1893.

Dolci, S., Williams, D. E., Ernst, M. K., Resnick, J. L., Brannan, C. I., Lock, L. F., Lyman, S. D., Boswell, H. S., and Donovan, P. J. (1991). Requirement for mast cell growth factor for primordial germ cell survival in culture. *Nature* **352**, 809–811.

Enders, G. C., Kahsai, T. Z., Lian, G., Funabiki, K., Killen, P. D., and Hudson, B. G. (1995). Developmental changes in seminiferous tubule extracellular matrix components of the mouse testis: Alpha 3(IV) collagen chain expressed at the initiation of spermatogenesis. *Biol. Reprod.* **53**, 1489–1499.

Godin, I., Deed, R., Cooke, J., Zsebo, K., Dexter, M., and Wylie, C. C. (1991). Effects of the *steel* gene product on mouse primordial germ cells in culture. *Nature* **352**, 202–203.

Hadley, M. A., Weeks, B. S., Kleinman, H. K., and Dym, M. (1990). Laminin promotes formation of cord-like structures by Sertoli cells *in vitro*. *Dev. Biol.* **140**, 318–327.

Huckins, C. (1963). Changes in gonocytes at the time of initiation of spermatogenesis in the rat. *Anat. Rec.* **145**, 145–243.

Jannini, E. A., Ulisse, S., Piersanti, D., Carosa, E., Muzi, P., Lazar, J., and D'Armiento, M. (1993). Early thyroid hormone treatment in rats increases testis size and germ cell number. *Endocrinology* **132**, 2726–2728.

Laslett, A. L., Li, L.-H., Jester, W. F., and Orth, J. M. (2000). Thyroid hormone (T3) downregulates NCAM expression and affects attachment of gonocytes in Sertoli cell-gonocyte cocultures. *Endocrinology* **141**, 1633–1641.

Li, H., Papadopoulos, V., Vidic, B., Dym, M., and Culty, M. (1997). Regulation of rat testis gonocyte proliferation by platelet-derived growth factor and estradiol: Identification of signaling mechanisms involved. *Endocrinology* **138**, 1289–1298.

Li, L.-H., Jester, W. F., and Orth, J. M. (1998). Expression of 140-kD neural cell adhesion molecule in developing testes *in vivo* and in long-term Sertoli cell-gonocyte cocultures. *J. Androl.* **19**, 365–373.

Lin, L. H., and DePhilip, R. M. (1996). Differential expression of placental (P)-cadherin in Sertoli cells and peritubular myoid cells during postnatal development of the mouse testis. *Anat. Rec.* **244**, 155–164.

Loveland, K., Schlatt, T., Sasaki, T., Chu, M. L., Timpl, R., and Dzidaek, M. (1998). Developmental changes in the basement membrane of the normal and hypothyroid postnatal rat testis: Segmental localization of fibulin-2 and fibronectin. *Biol. Reprod.* **58**, 1123–1130.

Mayerhofer, A,. Lahr, G., and Gratzl, M. (1991). Expression of the neural cell adhesion molecule in endocrine cells of the ovary. *Endocrinology* **129**, 792–800.

McCoshen, J. A., and McCallion, D. J. (1975). A study of the primordial germ cells during their migratory phase in steel mutant mice. *Experientia* **31**, 589–590.

McGuinness, M. P., and Orth, J. M. (1992a). Reinitiation of gonocyte mitosis and movement of gonocytes to the basement membrane in testes of newborn rats *in vivo* and *in vitro*. *Anat. Rec.* **233**, 527–537.

McGuinness, M. P., and Orth, J. M. (1992b). Gonocytes of male rats resume migratory activity postnatally. *Eur. J. Cell Biol.* **59**, 196–210.

Mintz, B., and Russell, E. (1957). Gene-induced embryological modifications of primordial germ cells in the mouse. *J. Exp. Zool.* **134,** 207–237.
Novi, A. M., and Saba, P. (1968). An electron microscopic study of the development of rat testis in the first 10 postnatal days. *Z. Zellforsch.* **86,** 313–326.
Nouwen, E. J., Dauwe, S., van der Biest, I., and De Broe, M. E. (1993). Stage- and segment-specific expression of cell-adhesion molecules N-CAM, A-CAM, and L-CAM in the kidney. *Kidney Int.* **44,** 147–158.
Orth, J. M. (1982). Proliferation of Sertoli cells in fetal and postnatal rats: A quantitative autoradiographic study. *Anat. Rec.* **203,** 485–492.
Orth, J. M. (1984). The role of follicle-stimulating hormone in controlling Sertoli cell proliferation in testes of fetal rats. *Endocrinology* **115,** 1248–1255.
Orth, J. M., Gunsalus, G. L., and Lamperti, A. A. (1988). Evidence from Sertoli cell-depleted rats indicates that spermatid number in adults depends on numbers of Sertoli cells produced during perinatal development. *Endocrinology* **122,** 787–794.
Orth, J. M., and Boehm, R. (1990a). Endorphin suppresses FSH-stimulated proliferation of isolated neonatal Sertoli cells by a pertussis toxin-sensitive mechanism. *Anat. Rec.* **226,** 32–327.
Orth, J. M., and Boehm, R. (1990b). Functional coupling of neonatal rat Sertoli cells and gonocytes in coculture. *Endocrinology* **127,** 2812–2820.
Orth, J. M., and McGuinness, M. P. (1991). Neonatal gonocytes co-cultured on a laminin-containing matrix resume mitosis and elongate. *Endocrinology* **129,** 1119–1121.
Orth, J. M., and Jester, W. F. (1995). NCAM mediates adhesion between gonocytes and Sertoli cells in cocultures from testes of neonatal rats. *J. Androl.* **16,** 389–399.
Orth, J. M., Jester, W. F., and Qiu, J. (1996). Gonocytes in testes of neonatal rats express the *c-kit* gene. *Mol. Reprod. Dev.* **45,** 123–131.
Orth, J. M., Qiu, J., Jester, W. F., and Pilder, S. (1997). Expression of the *c-kit* gene is critical for migration of neonatal rat gonocytes *in vitro. Biol. Reprod.* **57,** 676–683.
Packer, A. J., Besmer, P., and Bacharova, R. F. (1995). Kit ligand mediates survival of type A spermatogonia and dividing spermatocytes in postnatal mouse testes. *Mol. Reprod. Dev.* **42,** 303–310.
Rutishauser, U. (1991). Neural cell adhesion molecule and polysialic acid. *In* "Receptors for Extracellular Matrix" (J. McDonald and R. Mecham, eds.), pp. 132–156. Academic Press, New York.
Simorangkir, D. R., de Kretser, D. M., and Wreford, N. G. (1995). Increased numbers of Sertoli and germ cells in adult rat testes induced by synergistic action of transient neonatal hypothyroidism and neonatal hemicastration. *J. Reprod. Fertil.* **104,** 207–213.
Simorangkir, D. R., Wreford, N. G., and de Kretser, D. M. (1997). Impaired germ cell development in the testes of immature rats with neonatal hypothyroidism. *J. Androl.* **18,** 186–193.
Thiery, J.-P., Duband, J.-L., Rutishauser, U., and Edelman, G. M. (1982). Cell adhesion molecules in early chicken embryogenesis. *Proc. Natl. Acad. Sci. USA* **79,** 6737–6741.
Van Dissel-Emiliani, F. M., De Boer-Brower, M., and De Rooij, D. G. (1996). Effect of fibroblast growth factor-2 on Sertoli cells and gonocytes in coculture during the perinatal period. *Endocrinology* **137,** 647–654.
Van Haaster, L. H., de Jong, F. H., Docter, R., and de Rooij, D. G. (1993). High neonatal triiodothyronine levels reduce the period of Sertoli cell proliferation and accelerate tubular lumen formation in the rat testis, and increase serum inhibin levels. *Endocrinology* **133,** 755–760.
Yoshinaga, K., Nishikawa, S., Ogawa, M., Hayashi, S.-I., Kunisada, T., Fujomoto, T., and Nishikawa, S.-I. (1991). Role of *c-kit* in mouse spermatogenesis: Identification of spermatogonia as a specific site of *c-kit* expression. *Development* **113,** 689–699.

6
Attributes and Dynamics of the Endoplasmic Reticulum in Mammalian Eggs

Douglas Kline
Department of Biological Sciences
Kent State University
Kent, Ohio 44242

I. The Endoplasmic Reticulum (ER), Calcium, and Egg Activation
   A. Activation of the Egg at Fertilization
   B. Calcium Release from the ER
   C. Inositol 1,4,5-Trisphosphate (IP_3) Receptor Subtypes in Mammalian Eggs
   D. Calcium Waves and Oscillations
II. Development of the Calcium-Releasing System during Oocyte Maturation
   A. Sperm-Induced Calcium Release in Immature Mammalian Oocytes
   B. Increased Sensitivity to IP_3 Develops during Maturation of Oocytes
III. Arrangement and Reorganization of the ER
   A. Reorganization of the ER during Oocyte Maturation
   B. Redistribution and Increase in IP_3 Receptors
   C. The Role of ER Clusters in Generating Calcium Waves
   D. Stability of the ER during Calcium Oscillations
   E. Calcium and ER Dynamics
   References

The endoplasmic reticulum is a multifunctional continuous network of membrane-enclosed sacs and tubules that extends throughout the cell. The endoplasmic reticulum is the site of protein synthesis and assembly, as well as lipid and membrane synthesis. Additionally, the endoplasmic reticulum contains calcium pumps, intraluminal calcium storage proteins, and specific calcium-releasing channels. Thus, this membrane system plays a central role in intracellular signaling through the storage and release of calcium. At fertilization, the sperm triggers a large and dramatic release of calcium from the endoplasmic reticulum, which activates the egg to begin development. The ability of the egg to fully elevate calcium depends on biochemical and structural changes during oocyte maturation. The sensitivity of the calcium-releasing system increases and the endoplasmic reticulum is reorganized during maturation of the oocyte; together, these dynamic changes place a substantial calcium storage compartment just beneath the membrane, near the site of sperm–egg fusion. Localization of the calcium store may also contribute to the long-lasting calcium oscillations that are characteristic of mammalian fertilization. Examination of the endoplasmic reticulum in living eggs is leading to a better understanding of calcium release at fertilization. © 2000 Academic Press.

I. The Endoplasmic Reticulum (ER), Calcium, and Egg Activation

A. Activation of the Egg at Fertilization

Fertilization and early development of the zygote depend not only on the successful penetration of the egg by the sperm, but also on the ability of the egg to respond. Physiological and biochemical changes in the egg at fertilization are triggered by the fertilizing sperm. These changes activate the egg to begin development and the process is referred to as egg activation. Egg activation is dependent on the sperm-initiated release of calcium from the endoplasmic reticulum of the egg. This chapter focuses on the role of the ER in calcium signaling in the egg and the dramatic reorganization of the ER during oocyte maturation prior to fertilization.

Before considering the endoplasmic reticulum and calcium dynamics, it is necessary to define a few terms and outline the meiotic state of the female gamete at fertilization. Oogenesis, the differentiation of the female gamete to produce a mature egg or ovum, begins before meiosis is completed. In most species, growth and differentiation occur largely while the primary oocyte is arrested in prophase of the first meiotic division (prophase I). During this phase, the oocyte has a large intact nucleus (the germinal vesicle). While arrested in this state, the oocyte builds up the necessary materials needed for early embryonic development, including organelles, metabolic substrates, mRNA, and a variety of proteins. Meiosis must be completed in order to form a haploid female pronucleus, which can fuse with the fertilizing sperm pronucleus.

The timing of fertilization relative to the meiotic cycle varies depending on the species. For example, fertilization occurs after the completion of meiosis in sea urchins. However, in most species, the resumption of meiosis is initiated by hormones, the germinal vesicle membrane breaks down, and the egg is fertilized at a later stage of meiosis. Fertilization in some invertebrate species occurs at metaphase of the first meiotic division (metaphase I). In most vertebrates, including humans, meiosis proceeds as far as metaphase II, then is temporarily arrested a second time. During the period between prophase I and metaphase II, known as the period of oocyte maturation, molecular and physiological changes take place that are necessary for normal development after fertilization. Oocyte maturation transforms the immature, prophase I-arrested oocyte into the mature, metaphase II-arrested egg. The central consequence of egg activation in mammals and most other vertebrates is the release of the egg from metaphase II arrest.

Irrespective of species and regardless of when the egg is fertilized in the meiotic cycle, activation is triggered by calcium. Interaction of the fertilizing sperm with the egg triggers a dramatic increase in intracellular calcium (reviewed by Jaffe, 1985; Epel, 1989; Nuccitelli, 1991; Whitaker and Swann, 1993; Swann and Ozil, 1994; Schultz and Kopf, 1995; Miyazaki, 1995a; Ben-Yosef and Shalgi,

1998; Stricker, 1999). Three observations define the central role of calcium in egg activation. A sperm-induced rise in intracellular calcium is detected in eggs of species ranging from starfish to frog, humans, and even flowering plants (Stricker *et al.*, 1994; Busa and Nuccitelli, 1985; Taylor *et al.*, 1993; Digonnet *et al.*, 1997). Preventing the rise in calcium inhibits egg activation (e.g., Zucker and Steinhardt, 1978; Kline, 1988; Kline and Kline, 1992a). Artificially increasing calcium in the egg initiates activation in the absence of sperm (reviewed in Jaffe, 1985).

Beyond the resumption of meiosis, other calcium-dependent events are required for normal development. In some nonmammalian species, the earliest consequence of the calcium rise is the opening of calcium-gated ion channels in the membrane that produce the fertilization potential (electrical block to polyspermy). The calcium rise also triggers cortical granule exocytosis to establish a block to polyspermy. Modification of the zona pellucida after cortical granule exocytosis is the primary polyspermy-preventing mechanism in mammals. Other processes triggered by the calcium rise include the recruitment and translation of stored maternal mRNAs and other processes necessary for embryonic development.

B. Calcium Release from the ER

Injection of the lipophilic dye, DiI, and confocal imaging reveal that the endoplasmic reticulum is a large continuous network (Terasaki and Jaffe, 1991, 1993; Terasaki *et al.*, 1996). Demonstration that the ER stores and releases calcium in eggs was obtained from experiments in which eggs were centrifuged to localize the ER in one layer. Calcium measurements in the centrifuged sea urchin and frog egg demonstrated that calcium was released from the layer containing mostly ER (Eisen and Reynolds, 1985; Han and Nuccitelli, 1990). Calcium uptake and release have also been examined using preparations of the sea urchin egg cortex and microsomal vesicle preparations from whole sea urchin eggs (e.g., Oberdorf *et al.*, 1986). The bulk of these preparations is likely to be ER, although other organelles might be present. Direct evidence that calcium is sequestered and released from the ER was obtained by Terasaki and colleagues using an *in vitro* preparation of cortical ER (Terasaki *et al.*, 1991; Terasaki and Sardet, 1991).

Calcium serves as a key intracellular messenger in many cells that respond to biological signals. Transduction pathways linking the activating stimulus and calcium release from the ER are well defined in some cases. Inositol trisphosphate receptors are the primary calcium-releasing channels in the ER and play a central role in calcium signaling (reviewed by Miyazaki, 1995b). Ryanodine receptors may also be present in the ER membrane to mediate calcium release. Characterization of calcium signaling in eggs has relied largely on studies of marine organisms and amphibians. More recently, studies in mouse, hamster, pig, and other mammalian species, including humans, have shown that calcium release from the ER is mediated by the IP_3 receptor.

The second messenger, inositol 1,4,5-trisphosphate (IP_3), is generated following the stimulation of membrane receptors coupled to heterotrimeric G-proteins or by tyrosine kinase-linked receptors. IP_3 is the product of the hydrolysis of the membrane lipid, phosphatidylinositol 4,5-biphosphate (PIP_2), by phospholipase C (PLC). IP_3 binds with high affinity to the IP_3 receptor, which is the calcium channel that traverses the membrane of calcium storage compartments formed from the endoplasmic reticulum. The essential role for IP_3-mediated calcium release in eggs is supported by three observations: an increase in IP_3 at fertilization, injection of IP_3 releases calcium, and blocking IP_3 receptors inhibits calcium release.

The first indication that there might be inositol phospholipid turnover to produce IP_3 at fertilization was the detection of an increase in PIP_2 and its precursor, phosphatidylinositol monophosphate (PIP), within a minute after fertilization of the sea urchin egg (Turner et al., 1984). Other experiments later demonstrated that PIP_2 production also occurs in the frog egg (Snow et al., 1996). The natural increase in IP_3 concentration at fertilization is more difficult to demonstrate because large number of cells are needed for the usual assays, but experiments in frog eggs, which can be obtained in large numbers, have confirmed that IP_3 is produced at fertilization (Stith et al., 1993, 1994; Snow et al., 1996). Development of new fluorescent probes that could monitor IP_3 production or activity in single cells (e.g., Hirose et al., 1999) may permit measurement of IP_3 activity in mammalian eggs.

Microinjection of IP_3 causes calcium release in eggs of invertebrate and vertebrate species, including mammals (reviewed in Stricker, 1999). Calcium release in hamster and mouse eggs is completely inhibited by injection of an antibody against the IP_3 receptor (Miyazaki et al., 1992, 1993; Oda et al., 1999). Heparin and pentosan polysulfate (less specific inhibitors of the IP_3 pathway) block calcium release in frog (Nuccitelli et al., 1993) and sea urchin (Mohri et al., 1995) eggs, further suggesting that calcium release at fertilization is mediated in most species by IP_3. Inhibiting the enzyme PLC during fertilization of mouse eggs prevents the sperm-induced increase in intracellular calcium (Dupont et al., 1996), providing additional evidence for the central role of IP_3. Together with the evidence that a function-blocking antibody to the IP_3 receptor prevents calcium release and egg activation in mammalian eggs, the role for inositol phospholipid turnover and IP_3 production is quite strong.

The endoplasmic reticulum in eggs may contain both ryanodine and IP_3 receptors. Ryanodine receptors have been detected in sea urchin (McPherson et al., 1992), mouse (Ayabe et al., 1995), ascidian (Arnoult et al., 1997), and bovine (Yue et al., 1998) eggs. Ryanodine receptors were not detected by immunoblot in hamster eggs (Miyazaki et al., 1992). Agents that cause the opening of the ryanodine receptor, including ryanodine or the naturally occurring cyclic ADP ribose, initiate calcium release in sea urchin (reviewed by Shen, 1995), bovine

(Yue et al., 1995, 1998; but see also He et al., 1997), porcine (Machaty et al., 1997), and human (Sousa et al., 1996) eggs. However, the physiological significance of elevating calcium through the ryanodine receptor in eggs is unclear, given the central role of IP_3-mediated calcium release. While ryanodine receptor-mediated calcium release in the sea urchin egg is well documented, evidence suggests that the ryanodine receptor-mediated release plays only an accessory role at fertilization, as a variety of inhibitors against key components of ryanodine-mediated release do not block the sperm-induced increase in calcium (Lee and Shen, 1998). Likewise for mammals, while cyclic ADP ribose triggers calcium release in the eggs of some species, no response is seen in hamster or mouse (Whitaker and Swann, 1993; Kline and Kline, 1994) and inhibition of ryanodine receptors does not block the resumption of meiosis in mouse (Ayabe et al., 1995).

While the evidence for the generation and action of IP_3 at fertilization is strong, the mechanism leading to the production of IP_3 is not entirely known. This question goes beyond the scope of this review, but briefly, the research now revolves around determining how the sperm might activate egg PLC or introduce some other soluble or perinuclear calcium-releasing factor after sperm–egg fusion (Kimura et al., 1998; Swann and Parrington, 1999; Fissore et al., 1998; Wu et al., 1998). Several studies have focused on which of the three isoforms of PLC might be involved in IP_3 production at fertilization. PLCβ is activated by G-proteins, whereas PLCγ is activated by tyrosine kinases. Both PLCβ and PLCγ pathways are present in eggs of at least some species. Less is known about PLCδ, the third isoform of PLC; it is not known if it functions in eggs. Evidence now suggests that calcium release in echinoderm eggs (starfish and sea urchin) depends on the activation of PLCγ by a Src family kinase (Carroll et al., 1999; Giusti et al., 1999). Experiments suggest that PLCγ is not involved in mouse fertilization (Mehlmann et al., 1998), nor is BLCβ (Williams et al., 1998) unless these enzymes are activated in some unusual or atypical manner. One hypothesis is that a cytosolic sperm factor, which would be introduced into the egg after sperm–egg fusion, could be PLC (Jones et al., 1998a; Parrington et al., 1999). However, other candidates have been proposed, and the exact nature of the sperm-borne factor that induces calcium release in eggs remains to be determined.

C. Inositol 1,4,5-Trisphosphate Receptor Subtypes in Mammalian Eggs

IP_3 binds to one of three IP_3 receptor subtypes. The IP_3-gated calcium channel exists as a complex of four IP_3 receptor subunits, which may be homomeric or heteromeric in somatic cells. Mouse oocytes and eggs contain mRNA for all three receptor isoforms. Western analysis has shown that the type 1 IP_3 receptor is the predominate IP_3 receptor protein expressed in the mouse egg. The type 2 isoform has not been detected (Parrington et al., 1998; Fissore et al., 1999). The type 3 isoform was reported to be present in very low amounts (Fissore et al., 1999). Although the type 2 isoform was not detected by Western analysis, Fissore et al.

(1999) reported that the type 2 receptor could be found and localized by immunocytochemistry. However, the role of the type 2 receptor isoform is unclear because egg activation and calcium release studies have demonstrated that the calcium waves are mediated by the type 1 IP_3 receptor. Both egg activation and calcium release are completely prevented by the function-blocking monoclonal antibody 18A10 against the type 1 receptor isoform (Xu et al., 1994; Oda et al., 1999). The type 1 IP_3 receptor probably plays the major role in other species, including bovine (He et al., 1997) and human (Goud et al., 1999), although it will be necessary to determine if the other isoforms are present and if they make a significant contribution to calcium release.

D. Calcium Waves and Oscillations

At fertilization, intracellular calcium rises first at the site of sperm entry and then a wave of elevated calcium travels through the egg cytoplasm. The wave front passes through the cytoplasm at a velocity of 5–30 μm/sec, depending on the species and temperature (Stricker, 1999). Waves are propagated by the diffusion of calcium and, to some extent, by IP_3 diffusion, which promotes calcium release from the endoplasmic reticulum. The primary mechanism for wave propagation in mammals, and probably most species, is a form of calcium-induced calcium release in which calcium sensitizes IP_3-mediated calcium release (Miyazaki et al., 1993; Miyazaki, 1995a).

A single calcium wave is triggered at fertilization in echinoderm, cnidarian, fish, and frog eggs. In mammals and several other species, the first calcium wave is followed by repetitive transients or spikes (Fig. 1). Calcium oscillations have been reported to occur in fertilized eggs of all mammalian species investigated so far, including mouse (Kline and Kline, 1992a), hamster (Miyazaki et al., 1986), pig (Sun et al., 1992), bovine (Fissore et al., 1992), rat (Ben-Yosef et al., 1993), rabbit (Fissore and Robl, 1993), and human (Taylor et al., 1993). The repetitive calcium spikes in mammalian eggs may also be called oscillations. However, transients or spikes might be more descriptive terms because calcium does return to basal levels between spikes. Oscillations, however, can also occur on an elevated calcium plateau, which is a pattern of calcium release seen in some somatic cells.

The function of these repetitive spikes or oscillations following fertilization is not entirely known. It has been proposed that calcium oscillations may be necessary to fully inactivate the maturation promoting factor (MPF) and cytostatic factor (CSF) and to release the egg from metaphase arrest. MPF and CSF together maintain meiotic arrest at metaphase II, and calcium-dependent inactivation of MPF and CSF is necessary for resumption of the cell cycle (Collas et al., 1995; Zernicka-Goetz et al., 1995; Dupont, 1998; Lawrence et al., 1998). The function of calcium oscillations will be discussed in more detail in Section III,E.

Calcium oscillations in mouse and hamster eggs are dependent on extracellular calcium. Research indicates that the first calcium wave at fertilization depletes

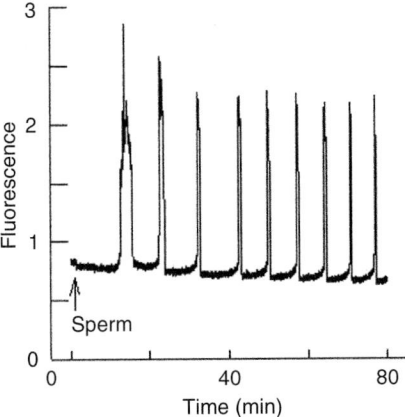

Fig. 1 Calcium transients after fertilization of the mouse egg by the sperm. At the arrow, sperm were added to the medium containing the egg that had previously been loaded with the calcium indicator Fura-2. An increase in intracellular calcium is represented by the increase in fluorescence [ratio of fluorescence for the 350/385-nm excitation wavelengths for Fura-2; see Mehlmann and Kline (1994) for details]. The recording was stopped at 80 min. Oscillations are known to continue in most eggs for several more hours (see text).

an intracellular calcium store and triggers capacitative calcium entry (Kline and Kline, 1992b; McGuinness et al., 1996). Capacitative calcium entry (calcium influx activated by depletion of a calcium store) exists in a number of cell types. Although not investigated in mouse, evidence shows that an IP_4-activated influx pathway exists in the hamster egg (Shirakawa and Miyazaki, 1995). Inositol 1,3,4,5-tetrakisphosphate (IP_4) is a product of IP_3 phosphorylation and is produced after a rise in IP_3. Injection of IP_4 triggers calcium influx, suggesting that calcium could be due to an IP_4-activated influx as well as a capacitative influx, which is mediated by an unknown mechanism.

The persistent influx of calcium is responsible for maintaining the repetitive calcium transients in the egg, which, in the mouse egg, propagate through the cytoplasm in a wave-like manner. Irrespective of the site of sperm entry, repetitive waves begin in the cortex of the vegetal hemisphere (Kline et al., 1999; Deguchi et al., 2000). Localization of ER in the cortex of the mature egg may be important for the capacitative influx and generation of calcium waves (discussed in Section III,C).

II. Development of the Calcium-Releasing System during Oocyte Maturation

Although the immature oocyte has the capacity for sperm–egg fusion, oocytes are normally fertilized after a period of oocyte maturation. During maturation, the oocyte gains the capacity to undergo normal responses to fertilization. For example, starfish oocytes develop the ability to produce a fertilization potential, to complete cortical granule exocytosis, and to decondense sperm nuclei (Miyazaki and Hirai, 1979; Hirai *et al.*, 1981). Likewise, preovulatory, immature mammalian oocytes are not competent to undergo complete cortical granule exocytosis; oocytes gain this capacity during maturation (reviewed by Ducibella, 1996, 1998). The inability of immature oocytes to produce calcium-dependent responses when fertilized led investigators to examine whether calcium release in oocytes is insufficient. Does the development of mechanisms to release calcium contribute to the capacity of mature eggs to produce the usual fertilization responses? The first answer to this question came from experiments with starfish oocytes and eggs.

Chiba *et al.* (1990) found that the amount of calcium released in fertilized starfish oocytes is less than in eggs and this difference was attributed to the reduced sensitivity of the IP_3-mediated calcium-releasing system in oocytes. To release the same amount of calcium, about 100 times more injected IP_3 was necessary in oocytes than in eggs. At very high IP_3 concentrations, the release of calcium in oocytes and eggs was comparable, suggesting that oocytes and eggs contain a similar sized store of calcium. It is the calcium-releasing system that develops during starfish oocyte maturation, not the amount of stored calcium.

A. Sperm-Induced Calcium Release in Immature Mammalian Oocytes

The initial calcium transient in fertilized mouse or hamster oocytes is much lower in amplitude and shorter in duration than in mature eggs (Fugiwara *et al.*, 1993; Mehlmann and Kline, 1994; Shiraishi *et al.*, 1995). For example, the amplitude of the first calcium transient in immature mouse oocytes is about one-third that of the mature egg, and the duration of the first calcium transient in oocytes is less than half as long as in mature eggs. Although oscillations occur in oocytes, the first calcium transient is likely to be responsible for the exocytosis of cortical granules (Kline and Stewart-Savage, 1994). The reduced calcium release can account, in part, for the inability of immature oocytes to undergo cortical granule exocytosis (although other calcium-independent factors may contribute to cortical granule exocytosis incompetence) (Ducibella, 1996, 1998).

The first sperm-induced calcium transient in hamster (Fujiwara *et al.*, 1993; Shiraishi *et al.*, 1995) and mouse oocytes (D. Kline, unpublished) is weakly propagated as a wave; a distinct wavefront is absent. Carroll and colleagues (1994) also found that calcium oscillations in immature oocytes triggered by a cytoplasmic extract of sperm are characteristically nonpropagating. A calcium wave is not seen and the calcium appears to increase in a homogeneous manner throughout the oocyte (Carroll *et al.*, 1994). The switch to distinct propagating Ca^{2+} waves is associated with oocyte maturation.

B. Increased Sensitivity to IP$_3$ Develops during Maturation of Oocytes

Like starfish oocytes, hamster and mouse oocytes develop an increased sensitivity to IP$_3$, which can account for greater calcium release in fertilized mature eggs. At any given IP$_3$ concentration, the peak calcium release in hamster oocytes was found to be about 35% of that in mature eggs injected with the same amount of IP$_3$. However, the response in oocytes could be increased to nearly that of mature eggs if the IP$_3$ concentration was increased (Fujiwara *et al.*, 1993).

IP$_3$-mediated calcium release in mouse oocytes is also less than in mature eggs, but can be increased to as much as that in eggs if the IP$_3$-dependent calcium-releasing system is sensitized by low concentrations of thimerosal (Mehlmann and Kline, 1994). At higher concentrations, thimerosal induces calcium oscillations by itself, probably because it sensitizes the calcium-releasing system, making it responsive to low levels of endogenous IP$_3$. Pretreatment of oocytes with 25 μ*M* thimerosal, which does not cause calcium oscillations, enhances calcium release by IP$_3$ in oocytes, bringing the level to that induced by the same IP$_3$ injection in untreated mature eggs.

Data indicate that hamster and mouse oocytes contain a substantial store of calcium in the endoplasmic reticulum, but this store is less responsive to IP$_3$ than the store in the mature egg. The human oocyte also appears to develop the calcium-releasing mechanism during maturation. Immature oocytes from small antral follicles are not very responsive to thimerosal, but mature eggs from luteinized follicles produce calcium oscillations (Herbert *et al.*, 1997). Similar development of the calcium release system is seen during both porcine (Machaty *et al.*, 1997) and bovine (He *et al.*, 1997) oocyte maturation. Data showing the development of calcium release mechanisms in mammalian and starfish oocytes led a number of investigators to examine mechanisms that might account for the increased sensitivity of the ER calcium store to IP$_3$.

III. Arrangement and Reorganization of the ER

Several early studies using fixed preparations of oocytes and eggs suggested that there is a reorganization of endoplasmic reticulum during oocyte maturation. In frogs, electron micrographs indicted that there is development of ER cisternae around cortical granules in the cortex during oocyte maturation (Gardiner and Grey, 1983; Charbonneau and Grey, 1984; Campanella *et al.*, 1984). A similar change was noted in mouse eggs in which membranous vesicles in the cortex increase in number following oocyte maturation (Ducibella *et al.*, 1988). More recently, the development of methods to label the ER with the fluorescent lipid probe, DiI, permitted further investigation of ER organization in living oocytes and eggs and has led to a greater understanding of the role of the ER in storing and releasing calcium. A dynamic reorganization of the ER, together with a redistribution and increase in IP$_3$ receptors, takes place during oocyte maturation.

A. Reorganization of the ER during Oocyte Maturation

The lipophilic dicarbocyanine dye, DiI [DiIC$_{18}$(3) or DiIC$_{16}$(3)] is used in conjunction with confocal microscopy to reveal the three-dimensional organization of the ER in oocytes or eggs. A saturated solution of DiI in soybean oil is injected into the egg. DiI transfers to any membrane in contact with the oil drop and diffuses through any continuous membranes (Terasaki and Jaffe, 1993). The validity of DiI labeling of the ER was confirmed in experiments with sea urchin eggs. The distribution of DiI is identical to an introduced green fluorescent protein (GPF) specifically targeted to the ER using a chimeric protein containing the KDEL ER retention sequence (Terasaki et al., 1996).

DiI was used to label the ER of the oocyte and mature egg of the mouse (Mehlmann et al., 1995). DiI labels the entire ER network within 10 min after injection of an 8 pl drop of oil (about 4% of the egg volume; Kline et al., 1999). The endoplasmic reticulum in the immature, prophase I-arrested mouse oocyte consists of a fine tubular network with large accumulations of membrane in the interior. The accumulations are distributed evenly throughout the oocyte cytoplasm, but are found less commonly in the cortex adjacent to the cell membrane (Fig. 2A). The ER is redistributed during oocyte maturation. After germinal vesicle breakdown (4–6 hr after release from meiotic arrest), while some ER accumulations remain in the interior, clusters of ER become localized in the cortex. By the completion of maturation, the metaphase II egg contains many large, brightly stained ER accumulations or clusters that are localized in the cortex immediately beneath the plasma membrane (Fig. 2B). These clusters, which are usually 1–2 μm in diameter, are absent in the deeper cytoplasm.

The ER clusters are localized in the same region of the egg that contains cortical granules and has overlying microvilli on the cell surface. ER accumulations are not found in the animal hemisphere cortex around the metaphase II chromosomes. The area around the spindle is also known as the amicrovillar or microvilli-free and cortical granule-free area, as microvilli and cortical granules are not found in the membrane and cortex in this region. Localization of ER clusters is thus similar to the polarized distribution of cortical granules and microvilli. It has been shown that this egg polarity extends to specific membrane proteins, including possible sperm-binding molecules (Evans et al., 2000).

The distribution and localization of DiI-labeled membranes in living eggs are consistent with observations of mouse egg membranes by electron microscopy. Electron micrographs reveal aggregates of small membranous vesicles in the cortex of the mature egg and these are absent in the cortex around the spindle (Ducibella et al., 1988; Fissore et al., 1999). Although not shown conclusively, these vesicle clusters, which are 1–3 μm in diameter, probably correspond to the DiI-labeled clusters in living eggs. If so, vesicles seen in the electron microscope images should be connected in some way to permit the diffusion of DiI. At high magnification, the vesicles appear more spherical than tubular, but at least some appear to be interconnected (Fissore et al., 1999). It is not known how much fix-

6. Endoplasmic Reticulum in Mammalian Eggs

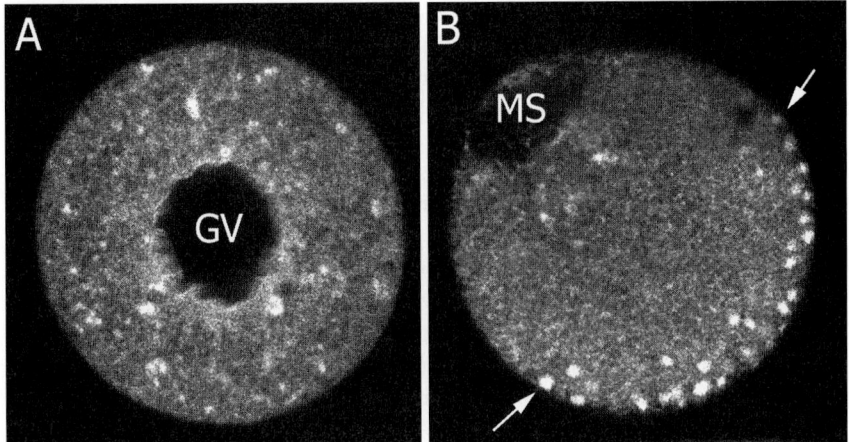

Fig. 2 The endoplasmic reticulum in equatorial confocal sections of a whole oocyte (A) and egg (B). The ER in each cell was labeled with the fluorescence dye DiI (for details, see Mehlmann et al., 1995). (A) Diffuse ER accumulations are found throughout the interior of the immature oocyte, in which the germinal vesicle (GV) appears dark. (B) ER accumulations are localized in the cortex of the mature egg and are absent in the hemisphere of the egg containing the meiotic spindle (MS). Arrows point to two clusters at the margin of the region that contains ER accumulations.

ation artifacts might alter the appearance of the clusters. Interconnections could be disrupted during fixation and processing for electron microscopy. Electron microscopy images also suggest that there is movement of ER to the cortex during oocyte maturation. The average number of vesicle clusters in the cortex of eggs increases to five times as many as in oocytes (Fissore et al., 1999).

The distribution of ER in hamster oocytes and eggs is somewhat different than in mouse, but the ER in both species is reorganized during maturation. In hamster, there is a transition from large irregular masses of ER in the oocyte to an ordered array of ER clusters in the cortex. The ER in immature hamster oocytes is densely distributed around the germinal vesicle and in large patches near the periphery of the cell. During oocyte maturation, these large masses divide up and disperse from subcortical regions into the cortex immediately beneath the plasma membrane forming smaller ER clusters. An ER-free zone surrounds the spindle and chromatin at one pole of the mature hamster egg. This region without ER is smaller in the mature hamster egg than the corresponding region in the mature mouse egg (Shiraishi et al., 1995). Preliminary examination of DiI-injected mature rat eggs (D. Kline, unpublished) indicates that cortical ER clusters in the rat egg closely resemble those seen in mouse eggs. Studies of the ER in rat oocytes or during oocyte maturation have not yet been done.

The physical clustering of ER in the cortex may contribute to the greater sensitivity of the mature egg to IP_3. Calcium release from the ER is dependent on the local concentration of both IP_3 and calcium. Calcium itself is known to enhance IP_3-induced calcium release (reviewed by Miyazaki, 1995b). Clustering of the ER may cause locally higher calcium concentrations, which would then result in greater IP_3 sensitivity. In addition, the PLC that generates IP_3 may be more concentrated in the plasma membrane; localization of the ER to the cortex would increase the overall sensitivity of the egg to IP_3. As discussed in Secton III,B., localization of the type 1 IP_3 receptor parallels the redistribution of ER.

Reorganization of the ER during oocyte maturation may also be common in nonmammalian species. Examination of fixed frog oocytes and eggs by electron microscopy shows a reorganization, particularly in the cortex. Furthermore, examination of living starfish oocytes using DiI revealed structural changes in the ER occurring during maturation. In starfish oocytes, there is a transition of the ER from a sheet-like form to a spherical form associated with yolk platelets. Moreover, there is usually an increase in ER movement during maturation. The ER of immature starfish oocytes is relatively stationary, but it begins to shift and move during maturation (Jaffe & Terasaki, 1994).

The mechanism by which the ER becomes reorganized during oocyte maturation is not known. In somatic cells, the distribution of the ER and its extension is closely associated with microtubules (Lee et al., 1989; Terasaki et al., 1986). Mouse oocytes and eggs contain cytoplasmic centrosomes with short microtubule arrays (Schatten et al., 1985, 1986; Maro et al., 1985). The relationship between these microtubule arrays and the ER has not been examined; however, there are far fewer centrosomes than ER clusters, so it is unclear how or if these microtubules participate in ER reorganization. Neither microtubules nor actin filaments appear to be required for the reorganization of the ER in starfish oocytes (Jaffe and Terasaki, 1994). The effects of microtubule- or actin-disrupting drugs, such as nocodazole or cytochalasin, on the reorganization of ER in mammalian eggs have not been examined. Cytochalasin D addition during mouse oocyte maturation prevents formation of the microvillar region and segregation of membrane markers of polarity (Evans et al., 2000), suggesting that actin microfilaments could be involved in establishing egg polarity. However, we do not know if the machinery that drives polarization of the egg membrane and cortical granules also operates in localization of the ER accumulations.

B. Redistribution and Increase in IP_3 Receptors

A dynamic redistribution of type 1 IP_3 receptors is associated with the changes in endoplasmic reticulum organization during oocyte maturation. For example, in mouse oocytes, immunofluorescent labeling of IP_3 receptors using a monoclonal antibody to the type 1 IP_3 receptor indicates that the type 1 IP_3 receptor is more abundant in the cortex of the mature egg than in the immature oocyte. Moreover, high magnification confocal microscopy of the egg cortex reveals that IP_3 receptors are localized in highly organized clusters that are about the same size as ER clusters. In contrast, IP_3 receptors in the oocyte are found in much smaller clumps or occasionally in more irregularly shaped patches (Mehlmann *et al.*, 1996).

A similar redistribution of IP_3 receptors was observed when mature human eggs were compared to immature germinal vesicle-stage oocytes. Prominent 2- to 5-μm-diameter IP_3 receptor clusters are found in the cortex of the mature egg (Goud *et al.*, 1999). In oocytes, IP_3 receptor staining is diffuse and cortical labelling is sparse compared to eggs. In hamster also, Shiraishi *et al.* (1995) found a large increase in IP_3 receptor labeling in the cortex, as well as the interior cytoplasm of mature eggs compared to immature oocytes. An increase in cortical IP_3 receptors may be a common feature of maturation in nonmammalian species also. For example, IP_3 receptors become concentrated in patches in the cortex after maturation of *Xenopus* oocytes (Parys *et al.*, 1994; Kume *et al.*, 1993, 1997).

Simultaneous imaging of the ER and IP_3 receptors during oocyte maturation has not been done. It is easier and preferable to label the ER in living cells, whereas IP_3 receptors are usually labeled in separate experiments after fixation. Some caution is appropriate in interpreting data from fixed cells because the ER can be disrupted by some fixation procedures. Comparable views of a living egg labeled with DiI and a different fixed egg labeled with an antibody to the type 1 IP_3 receptor are shown in Fig. 3. IP_3 receptors are clustered in the cortex. It may soon be possible to label IP_3 receptors and ER at the same time in living cells using optical probes. It would also be useful to compare IP_3 receptor distribution and ER localization by electron microscopy. However, thus far, it has not been possible to localize IP_3 receptors in eggs using electron microscopy (Fissore *et al.*, 1999). Nevertheless, when examined at the level of the light microscope, the

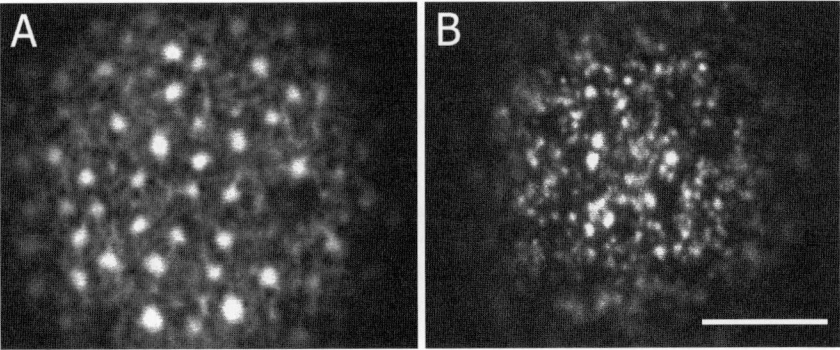

Fig. 3 Endoplasmic reticulum (A) and IP_3 receptors (B) in a portion of the cortex of the mature mouse egg imaged at high magnification with a confocal microscope. (A) The pattern of staining in the cortex of a living egg after DiI injection. (B) Immunolocalization of the type 1 IP_3 receptor in another fixed egg. IP_3 receptors are found in clusters in the cortex in a pattern similar to ER clusters (for details, see Mehlmann *et al.*, 1996). Magnification is the same for both photographs; the bar in B represents 10 μm.

redistribution of IP_3 receptors occurs in parallel to the redistribution of ER during oocyte maturation. In addition to the change in staining pattern, there is generally greater immunohistochemical or immunofluorescence staining in mature eggs than in oocytes, suggesting that there is more IP_3 receptor in eggs than oocytes. This was confirmed by a quantitative Western blot analysis of IP_3 receptors in oocytes and eggs.

Three separate studies with mice have shown that the immunoreactive mass of the type 1 IP_3 receptor is 1.8- to 1.9-fold greater in mature eggs compared to immature oocytes. (Mehlmann *et al.*, 1996; Parrington *et al.*, 1998; Fissore *et al.*, 1999). A smaller but significant increase is also seen following the maturation of bovine oocytes (He *et al.*, 1997). The increase in IP_3 receptor immunoreactive mass suggests that the number of IP_3 receptors in the egg is greater than in the oocyte. More IP_3 receptors, together with their localization in cortical ER clusters, could contribute to the greater calcium response in eggs to IP_3 injections and to sperm.

C. The Role of ER Clusters in Generating Calcium Waves

Sperm may fuse with the egg membrane anywhere on the surface with microvilli overlying cortical granules and ER clusters. The first calcium wave begins in the cortex at the site of sperm–egg fusion (Miyazaki *et al.*, 1986, 1993; Lawrence *et al.*, 1997; Jones *et al.*, 1998b). Likewise, injection of a cytosolic sperm extract or IP_3 induces calcium waves when injected in the cortical region. Moreover, in a deft series of microinjection experiments, Oda and colleagues (1999) demonstrated that the mouse cortex is substantially more sensitive to injected IP_3 and sperm extract than the interior. Localization of the calcium storage and releasing system (ER clusters and IP_3 receptors) in the cortex immediately beneath the site of sperm–egg fusion provides a means for ensuring the rapid generation of a calcium wave following sperm–egg fusion.

Similar clusters of ER containing IP_3 receptors are found in *Xenopus* oocytes. Confocal imaging reveals that calcium release in response to weakly released caged IP_3 or injection of a poorly metabolized IP_3 derivative evokes calcium puffs, which are 1–2 μm in diameter and localized in the cortex. Calcium spreads radially from these sites (Yao *et al.*, 1995; Callamaras and Parker, 1999). High-resolution studies of the spatial and temporal release of calcium in mammalian eggs are yet to be done, but the structure of the ER suggests that a significant part of the calcium release at the initiation of the calcium wave might be from discrete foci, as in the frog oocyte.

In mouse, the later calcium oscillations do not always begin at the site of sperm–egg fusion. It is possible to label the meiotic egg chromosomes and the fertilizing sperm with the DNA stain, Hoechst, and simultaneously image calcium to determine the orientation of the secondary waves. The secondary waves, in contrast to the first sperm-induced wave, begin in the hemisphere opposite the meiotic spindle (180° from the spindle). Thus, the later calcium transients in the mouse egg originate from a pacemaker-like region in the vegetal hemisphere (Kline et al., 1999; Deguchi et al., 2000).

Localization of cortical ER clusters and IP_3 receptors, together with the observation that calcium waves begin in the vegetal hemisphere, may provide some insight into the mechanics of wave generation. Calcium oscillations in the mouse egg are most likely the consequence of the opening of a capacitative calcium influx pathway (Kline and Kline, 1992b; McGuinness et al., 1996). Although the signal that links store depletion and capacitative calcium influx is not known, localization of the storage compartment near the cell membrane may be important. At least one model for capacitative calcium influx predicts a close association between the ER and the cell membrane (Jaconi et al., 1997).

We do not know if calcium entry in the fertilized mouse egg is localized to the membrane overlying ER accumulations, but the origin of calcium waves in the vegetal hemisphere cortex could be due to the localized influx and filling of calcium stores, which periodically empty, triggering repetitive waves. Capacitative influx appears to be triggered in fertilized immature oocytes, as oscillations occur (Mehlmann et al., 1996), but waves are weakly propagated and do not appear to originate from a particular region. The change to propagated waves after oocyte maturation may be related to the modification of the ER during oocyte maturation.

D. Stability of the ER during Calcium Oscillations

The structure of the ER during fertilization is of interest because, just as the reorganization during maturation seems to influence calcium release, changes after fertilization may also alter the calcium storage system, affecting both calcium release and wave propagation. Examination of the ER during fertilization could lead to a better understanding of how calcium waves are initiated and propagated in eggs. In a series of experiments, Jaffe and Terasaki examined ER structure during the fertilization of sea urchin and starfish eggs; they discovered a striking change in the appearance of the ER. The ER becomes more finely divided and there is a transient disruption of membrane continuity at fertilization.

The first evidence for changes in the ER during fertilization was obtained in experiments in which the appearance of DiI-labeled ER in living sea urchin and starfish eggs was examined with high-resolution confocal microscopy (Terasaki and Jaffe, 1991; Jaffe and Terasaki, 1994). At fertilization, a change in ER structure proceeds as a wave through the egg. Membrane sheets become more finely divided. This change is transient and is coincident with the rise and fall of cytoplasmic calcium. The prefertilization ER structure returns at about the time the calcium rise ends. In addition to the structural changes visualized by confocal microscopy, a transient fragmentation of the ER is indicated by reduced DiI diffusion. The rate of spreading is much slower when DiI in injected just after fertilization compared to eggs before fertilization or to eggs 10 min after fertilization. This observation suggests a marked decrease in membrane continuity. More recently, fluorescence recovery after photobleaching (FRAP) was used to confirm this observation that ER membranes in echinoderm eggs become discontinuous after fertilization (Terasaki et al., 1996).

The ER is a continuous membrane system, and a fluorescent probe in the membranes or in ER lumen is free to diffuse into a small region of the ER after the probe in that area is bleached with high-intensity laser illumination. Membrane discontinuities are reflected in poor fluorescence recovery after photobleaching. When a region of the unfertilized egg labeled with DiI, or the luminal protein GFP-KDEL, is bleached, recovery is rapid. However, Terasaki and colleagues (1996) showed that the time for recovery after photobleaching was much longer in the fertilized starfish egg. Recovery of fluorescence in DiI-labeled eggs was about five times longer in the fertilized egg than in the unfertilized egg or the egg 20 min after fertilization. The longer recovery time is consistent with a transient fragmentation of the ER, which would reduce diffusion pathways for the dye. The functional significance of this disruption of ER membranes at fertilization of echinoderm eggs is unknown; however, it is clearly associated with the rise in intracellular calcium as the ER changes are coincident with the rise and fall of calcium within the egg and can be mimicked by the injection of IP_3. Disruption of the ER could be caused by the rise in intracellular calcium. A large transient increase in intracellular calcium can fragment the ER in somatic cells (Subramanian and Meyer, 1997). Once the ER is disrupted, the calcium-releasing system may behave differently.

Taken together, these experiments provide strong evidence for the disruption of ER membranes at fertilization in echinoderm eggs. Based on these results, we examined the organization and dynamics of the ER in the mouse egg at fertilization and during the repetitive calcium transients. Direct imaging of the ER in DiI-labeled mouse eggs revealed little change in the cortical ER clusters or the fine ER surrounding them during fertilization and during calcium oscillations following fertilization (Kline et al., 1999). Furthermore, there was substantial fluorescence recovery in ER clusters after photobleaching, indicating that the ER in mouse eggs remains continuous during fertilization. In contrast to the transient disruption and reorganization of the ER that occur in the echinoderm egg, the ER in the mouse egg remains intact without displaying significant discontinuity through at least seven calcium transients (Kline et al., 1999).

Further work is needed to determine the significance of the finding that ER dynamics in echinoderm eggs and mouse eggs are strikingly different. However, the use of fluorescent dyes to label the ER in living cells and fluorescent indicators for measuring calcium has provided some data that allow us to speculate on the relationship among ER structure, calcium release, and the requirements for egg activation. One of the most striking differences among species is that, in some species, a single calcium transient occurs at fertilization; in these cases, one calcium rise is sufficient for complete egg activation. In contrast, calcium oscillations occur after fertilization in some species. Why are oscillations present in some species? How might the ER organization influence the pattern of calcium release? In order to examine the relationship between the ER and calcium oscillations, the author would like to compare calcium release in several animal groups and then examine the ER dynamics, which may have a role in affecting these calcium changes.

E. Calcium and ER Dynamics

1. Single and Multiple Calcium Transients at Fertilization

A number of investigators have speculated on the reason why some species produce calcium oscillations; it has been proposed that the occurrence of oscillations is related to the point in the meiotic cycle at which fertilization occurs as well as the length of the cell cycle (Swann and Ozil, 1994; Jones, 1998; Stricker, 1999; Kline et al., 1999). The most striking case is for mammals. Calcium oscillations have been reported to occur in fertilized eggs of all mammalian species investigated so far, including mouse, hamster, pig, cow, rat, rabbit, and human (for references, see Table I). In these groups, fertilization occurs at metaphase II and meiosis is completed several hours later when male and female pronuclei form (about 6–8 hr after fertilization in mouse).

Calcium oscillations in mammals occur at varying frequencies, depending on the species, but regardless of the frequency, they persist for many hours. The length of the oscillatory period has not been determined for all species, but investigators all report that oscillations last for several hours (often as long as recordings are made, e.g., Fig. 1). In mouse, oscillations usually persist for about 4 hr, through formation of the second polar body and continuing until just before pronuclear formation (Jones et al., 1995). The cessation of calcium oscillations may be associated with the downregulation of type 1 IP_3 receptors, which is known to occur after fertilization of the mouse egg (Parrington et al., 1998). Regulation of calcium release in the cell may also be dependent on factors or events that regulate cell cycle timing. Calcium oscillations persist for as long as 18 hr if the egg is arrested in the metaphase state with colcemid (Jones et al., 1995). Thus, some part of the cell cycle control machinery appears to feed back on the calcium-releasing system to regulate its sensitivity.

Mammalian eggs are fertilized internally and there may be a fairly long period between ovulation and the fertilization of the egg, which is arrested at metaphase II. In contrast, nonmammalian eggs are usually fertilized very soon after ovulation. This difference has led to speculation (e.g., Jones, 1998) that the cell cycle-arresting system may be substantially stronger in mammals than in other species in which fertilization occurs rapidly. A relatively high MPF and CSF activity in mammals may be required to prevent premature activation and the development of blocks to polyspermy prior to sperm–egg fusion. A single, short calcium transient might not be enough to overcome cell cycle arrest, and a long period of calcium signaling may be required in mammals to release the egg from metaphase arrest. This is not a long, sustained elevation of calcium because persistent elevation of calcium can damage cells. Rather, there are transient, repetitive calcium spikes, with calcium returning to the resting level between spikes. This rationale is supported by several lines of evidence.

The idea that calcium oscillations are necessary for the activation of mammalian eggs is suggested by several observations and experiments. Interestingly, the cell cycle-arresting system in mouse becomes less able to maintain arrest in eggs held for longer periods than normal. With a longer time after ovulation, eggs become more sensitive to artificial activation or may even activate spontaneously. Fertilization normally occurs about 12–14 hr after ovulation; at this time, most eggs are not readily activated by treatments that produce a single calcium transient, including calcium ionophore, ethanol treatment, calcium injection, or IP_3 injection. With age, eggs are more easily artificially activated by a single calcium transient, which correlates with a partial decrease in resting MPF activity in older eggs (Xu et al., 1997). These examinations of artificial activation suggest that a single calcium transient, whether induced by some artificial agent or by sperm, is not sufficient to activate the freshly ovulated egg.

The strength of the cell cycle-arresting system, indicated by the high MPF activity in freshly ovulated eggs, may require more sustained calcium signaling in the form of repeated transients. This idea is supported by indirect experiments in rabbit eggs that suggest that repeated calcium spikes may be necessary to maintain MPF inactivation over long periods (Collas et al., 1995). It was shown that MPF activity (indicated by H1 kinase activity) decreases following a single calcium increase induced by electrical stimulation, but MPF activity returned after the first calcium spike and additional electrical pulses were required to maintain MPF at low activity. These observations are consistent with earlier studies suggesting that the artificial activation of rabbit eggs was more successful if eggs were activated by repetitive electrical pulses to elevate calcium in a pattern that mimicked fertilization (Ozil, 1990, 1998).

More direct evidence to show that calcium oscillations are required has been somewhat difficult to obtain. To demonstrate that only one calcium transient is sufficient, all but the first must be blocked. It must also be shown that only the calcium-releasing system is blocked and that the method used does not block the cell cycle regulatory system nonspecifically at some downstream site. Using a heavy metal chelator in combination with a thiol-reducing agent to chelate intracellular calcium after one to several transients, Lawrence et al. (1998) showed that at least nine calcium oscillations may be the minimum necessary to ensure

6. Endoplasmic Reticulum in Mammalian Eggs

complete activation and pronuclear formation in all fertilized mouse eggs. In these experiments, the percentage of eggs activated (male and female pronuclei formed) decreased with fewer calcium transients.

Calcium oscillations occur after fertilization in eggs of ascidians, some bivalve molluscs, an annelid, and a nemertean (see Table I for references and reviews by Sardet *et al.*, 1998; Stricker, 1999). Like mammals, a relatively long period of calcium signaling may be required to drive the cell through meiosis. Fertilization in these species is at metaphase I and the calcium transients usually last through the formation of the second polar body. The duration of time for calcium oscillations is between 30 and 75 min with the exception of eggs of the annelid worm *Chaetopterus* in which calcium transients last for about 10 min.

In contrast, there is only a single calcium transient at fertilization in frog, fish, sea urchin, starfish, cindarian, and echiuran eggs (Table I). Like mammals, fish and frog eggs are fertilized at metaphase II; however, the transition to interphase is much shorter. For example, the frog egg forms the female pronucleus within 20 min after fertilization (Stewart-Savage and Grey, 1982); a single calcium increase is all that is needed to release the egg from meiotic arrest and move the egg to interphase. Hydrozoan jellyfish eggs (phylum Cnidaria) and sea urchin eggs are fertilized in the pronuclear stage, after the completion of meiosis. The calcium increase in these eggs is quite short, perhaps because the calcium signal, although needed for egg activation, is not needed for the resumption of meiosis. The *Urechis* egg (phylum Echiura) is fertilized in prophase I (germinal vesicle stage). The calcium transient is usually a single transient, although in some cases, low-frequency oscillations may occur on or after the calcium plateau (Stephano and Gould, 1997). A single calcium pulse may be sufficient in *Urechis* to release the egg from prophase I arrest and no further meiotic arrest occurs in this species. Starfish eggs are similar; fertilization at metaphase I results in a single calcium transient that is over by the time first polar body formation occurs. A long calcium signal may not be necessary to complete the meiotic cell cycle. Maturing starfish oocytes will progress through the meiotic cell cycle without fertilization because there is no natural checkpoint after germinal versicle breakdown.

To conclude this species comparison, some mention should be made regarding several other species of animals that are not included in Table I. These include a group of animals in which there is a single calcium stimulus for activation, but the calcium increase is not initiated by sperm, but by the spawning medium. The marine shrimp, *Sicyonia ingentis* (Lindsay *et al.*, 1992), the prawn, *Palaemon serratus* (Goudeau and Goudeau, 1996), and the zebrafish, *Danio rerio* (Lee *et al.*, 1999), are activated at spawning. A prolonged calcium signal is not necessary to overcome meiotic arrest. For example, the calcium pulse in zebrafish eggs last only about 7 min. At the beginning of this calcium transient, sperm have a very narrow window (<30 sec) to fertilize the egg before polyspermy preventive mechanisms are set in place as a consequence of the calcium rise.

TABLE I Calcium Dynamics and ER Properties at Fertilization

Animal	Meiotic stage for fertilization[a]	Calcium oscillations	Approximate duration of calcium elevation (min)[b]	Change and disruption of ER at fertilization[c]
Mammal (mouse, cow, hamster, pig, rabbit, rat, human)[d]	MII	Yes	4hr[e]	No[f]
Nemertea (Cerebratulus)	MI	Yes	75[g]	No[h]
Molluscs	MI	Yes	30[i]	—
Ascidians	MI	Yes	30[j]	No[k]
Annelid (Chaetopterus)	MI	Yes	10[l]	—
Echiuran (Urechis caupo)	GV	No[m]	5–7[m]	—
Fish (Oryzias latipes)	MII	No	10–15[n]	—
Frog (Xenopus laevis)	MII	No	15[o]	Yes[p]
Cnidarian (Mitrocomella, Phialidium)	PN	No	2–4[q]	—
Echinoderms				
Sea urchin	PN	No	3–5[r]	Yes[s]
Starfish	MI	No	10–15[t]	Yes[u]

[a]GV, germinal vesicle stage (prophase I arrest); MI, metaphase I arrest; MII metaphase II arrest; PN, pronuclear egg.

[b]Duration of the period during which calcium oscillations occur or duration of a single transient (min, except for mammals).

[c]Evidence obtained from confocal imaging of living eggs indicating whether the ER is disrupted or becomes transiently discontinuous at fertilization. A dash indicates that measurements have not yet been made in any species in those groups.

[d]Mouse (Cuthbertson and Cobbold, 1985; Kline and Kline, 1992a), hamster (Miyazaki et al., 1986), pig (Sun et al., 1992), bovine (Fissore et al., 1992), rat (Ben-Yosef et al., 1993), rabbit (Fissore and Robl, 1993), and human (Taylor et al., 1993; Tesarik and Sousa, 1994).

[e]Data from mouse (Jones et al., 1995).

[f]From Kline et al. (1999).

[g]From Stricker (1996).

[h]From Stricker et al. (1998).

[i]From Deguchi and Osanai (1994b) (oscillations occur in several species, a single transient was reported in one species, see Deguchi and Osanai, 1994a).

[j]From Speksnijder et al. (1989, 1990); Sardet et al., (1998); and McDougall and Sardet (1995).

[k]From Speksnijder et al. (1993).

[l]From Eckberg and Miller (1995).

[m]From Stephano and Gould (1997) (calcium oscillations on an elevated plateau may occur in some eggs, see text).

[n]From Ridgway et al. (1977) and Gilkey et al. (1978).

[o]From Busa and Nuccitelli (1985).

[p]M. Terasaki, personal communication.

[q]From Freeman and Ridgway (1993).

[r]From Stricker et al. (1992) and Shen and Buck (1993).

[s]From Terasaki and Jaffe (1991) and Jaffe and Terasaki (1993).

[t]From Eisen and Reynolds (1984), Stricker et al. (1994), and Stricker (1995).

[u]From Jaffe and Terasaki (1994) and Terasaki et al. (1996).

6. Endoplasmic Reticulum in Mammalian Eggs   145

In summary, two calcium-signaling strategies emerge. The calcium signal may need to be long lasting for those species fertilized at metaphase I and for mammals fertilized at metaphase II. Rather than prolonged calcium elevation, the signal is provided by repetitive calcium transients. Oscillations persist, usually through the completion of second polar body formation, and somewhat longer in mammals. The calcium signal in mammals may be the longest because of a robust meiotic-arresting system and a longer transition to interphase. The calcium signal need not be long lasting in eggs fertilized at the pronuclear stage, in eggs fertilized at the GV or MI stage that have no further cell cycle arrest, in MII eggs that have a very short transition to interphase, or in eggs that are activated by spawning. Eggs that only require a short period of calcium stimulation (e.g., 15 min or less) produce a single calcium transient (one exception may be the *Chaetopterus* egg in which oscillations are present, but persist for only about 10 min). What then is the influence of ER structure on these calcium-signaling strategies?

2. ER Dynamics at Fertilization

A stable organization of the ER may be necessary for generating or permitting long-lasting calcium oscillations, which, in turn, may be important for activation of the egg in some species. In those species that produce a single calcium transient, a stable ER organization is not essential and a transient disruption of the ER might even prevent additional calcium waves. These ideas are supported by several observations. The ER structure has been examined in eggs of some of the species that produce oscillations. There is no immediate change in ER structure during fertilization of ascidian, nemertean, or mouse eggs (Table I). The ER in ascidians eggs is redistributed after fertilization, forming an ER-rich area at the contraction pole in the vegetal hemisphere. Based on morphological observations, the ER clusters in the ascidian egg (referred to as ER microdomains), remain intact throughout the period of calcium oscillations, and the vegetal ER-rich region serves as the pacemaker region for calcium waves (Speksnijder *et al.*, 1993). Cortical ER accumulations in the nemertean egg are not as strikingly polarized as those in ascidian or mouse eggs. Except for a small region associated with the meiotic spindle, the distribution of ER clusters in the animal and vegetal hemisphere of the *Cerebratulus* egg is similar. However, later calcium waves begin in the cortex of the vegetal hemisphere, again suggesting that the cortical ER serves a pacemaker function (Stricker *et al.*, 1998). Fluorescence photobleaching experiments have not been done with ascidian or nemertean eggs to confirm that the ER remains continuous after fertilization; however, morphological observations indicate that, like mouse, the ER clusters are relatively stable.

The cortical ER forms a pacemaker region in the mouse egg that remains unaltered through at least seven calcium transients and the FRAP experiments indicate a persistent continuity (Kline *et al.*, 1999). Interestingly, ER accumulations in the nemertean egg, although showing little change during the first several calcium waves, begin to disappear about 40 min after fertilization. ER clusters in these eggs are not present by the time the calcium oscillations end (Stricker et al., 1998). We do not yet know if the ER clusters remain intact after the end of the calcium oscillations in mammalian eggs or if, as in the nemertean egg, they disappear at about the same time the oscillations cease.

When the ER has been studied in eggs that have only a single calcium rise, the ER is altered (Table I). In sea urchin (Terasaki and Jaffe, 1991; Jaffe and Terasaki, 1993) and starfish (Jaffe and Terasaki, 1994; Terasaki et al., 1996), eggs, striking transient changes in ER appearance correspond to periods of discontinuity. Frog eggs may also fit this pattern; observations indicate that the ER structure changes drastically during artificial activation (M. Terasaki, personal communication). These observations suggest that transient changes in the ER may somehow prevent oscillations in these species. It would be interesting to examine eggs of other species that do not produce calcium oscillations to learn if the ER also becomes discontinuous in those species.

3. Concluding Comments

As described in this chapter, the development of methods to examine the fine structure and the dynamics of the endoplasmic reticulum in living cells has provided opportunities to discover how organization of the ER influences calcium signaling and egg activation. The ER is reorganized during oocyte maturation to localize calcium storage compartments near the membrane and site of sperm–egg fusion. This reorganization, together with an increase in the number of calcium-releasing channels in the ER membrane, could account for greater calcium release in the mature egg. In addition, in some species, including mammals, the cortical localization of ER may set up calcium pacemakers in which calcium waves are periodically initiated following fertilization. In other species, the ER may be transiently disrupted at fertilization after a single calcium increase, thereby inhibiting further calcium release. Thus, the organization and dynamic changes in the ER may have a significant effect on calcium regulation.

Additional experiments are needed to test the importance of morphological changes during oocyte maturation and the significance of the transient disruption of the ER at fertilization. For example, it might be possible to prevent cortical localization during oocyte maturation to clarify its role in generating calcium waves in the mature egg. Examination of the ER in other species at fertilization would provide additional support for the idea that transient disruption of the ER has a role in preventing calcium release and provide insights into the mechanism by which this occurs.

The polarized organization of the ER in the mouse egg gives the entire egg and precleavage embryo a polarity that could be important in later development. There is some question about how the animal–vegetal polarity of the egg is reflected in embryonic polarity or other developmental events (reviewed in Gardner, 1999a,b), but it is possible that the polarized distribution of the ER is important. Calcium changes mediated by PLC may regulate aspects of preimplantation development (Stachecki and Armant, 1996a,b; Pey *et al.*, 1998; Wang *et al.*, 1998). We do not yet know if the cortical ER accumulations in the egg are retained. The downregulation of IP_3 receptors suggests that some changes in the mammalian egg ER may ultimately occur after fertilization. The structure of the ER has not been examined during pronuclear formation and we do not yet know what becomes of cortical ER accumulations during cleavage. Preliminary observations (D. Kline, unpublished observation) of two- to eight-cell embryos reveal that these early cleavage-stage cells have ER clusters similar to those in the mature egg. However, the structure and possible localization of ER accumulations in cleavage-stage cells and in cells in compaction and cavitation stages have not been determined. Continued investigations of the endoplasmic reticulum in oocytes, eggs, and early embryos should provide further insights to increase our understanding of early development.

Acknowledgments

The author thanks Lisa Mehlmann, Mark Terasaki, and Laurinda Jaffe for substantial and important contributions to the studies of mouse oocyte and egg ER. Work in the author's laboratory was supported by the NIH.

References

Arnoult, C., Albrieux, M., Antoine, A. F., Grunwald, D., Marty, I., and Villaz, M. (1997). A ryanodine-sensitive calcium store in ascidian eggs monitored by whole-cell patch-clamp recordings. *Cell Calcium* **21,** 93–101.

Ayabe, T., Kopf, G. S., and Schultz, R. M. (1995). Regulation of mouse egg activation: Presence of ryanodine receptors and effects of microinjected ryanodine and cyclic ADP ribose on uninseminated and inseminated eggs. *Development* **121,** 2233–2244.

Ben-Yosef, D., Oron, Y., and Shalgi, R. (1993). Prolonged, repetitive calcium transients in rat oocytes fertilized *in vitro* and *in vivo*. *FEBS Lett.* **331,** 239–242.

Ben-Yosef, D., and Shalgi, R. (1998). Early ionic events in activation of the mammalian egg. *Rev. Reprod.* **3,** 96–103.

Busa, W. B., and Nuccitelli, R. (1985). An elevated free cytosolic Ca^{2+} wave follows fertilization in eggs of the frog, *Xenopus laevis*. *J. Cell Biol.* **100,** 1325–1329.

Callamaras, N., and Parker, I. (1999). Radial localization of inositol 1,4,5-trisphosphate-sensitive Ca^{2+} release sites in *Xenopus* oocytes resolved by axial confocal linescan imaging. *J. Gen. Physiol.* **113,** 199–213.

Campanella, C., Andreuccetti, P., Taddei, C., and Talevi, R. (1984). The modifications of cortical endoplasmic reticulum during in vitro maturation of *Xenopus laevis* oocytes and its involvement in cortical granule exocytosis. *J. Exp. Zool.* **229,** 283–293.

Carroll, D. J., Albay, D. T., Terasaki, M., Jaffe, L. A., and Foltz, K. R. (1999). Identification of PLCγ-dependent and -independent events during fertilization of sea urchin eggs. *Dev. Biol.* **206,** 232–247.

Carroll, J., Swann, K., Whittingham, D., and Whitaker, M. (1994). Spatiotemporal dynamics of intracellular $[Ca^{2+}]_i$ oscillations during the growth and meiotic maturation of mouse oocytes. *Development* **120,** 3507–3517.

Charbonneau, M., and Grey, R. D. (1984). The onset of activation responsiveness during maturation coincides with the formation of the cortical endoplasmic reticulum in oocytes of *Xenopus laevis*. *Dev. Biol.* **102,** 90–97.

Chiba, K., Kado, R. T., and Jaffe, L. A. (1990). Development of calcium release mechanisms during starfish oocyte maturation. *Dev. Biol.* **140,** 300–306.

Collas, P., Chang, T., Long, C., and Robl, J. M. (1995). Inactivation of histone H1 kinase by Ca^{2+} in rabbit oocytes. *Mol. Reprod. Dev.* **40,** 253–258.

Cuthbertson, K. S., and Cobbold, P. H. (1985). Phorbol ester and sperm activate mouse oocytes by inducing sustained oscillations in cell Ca^{2+}. *Nature* **316,** 541–542.

Deguchi, R., and Osanai, K. (1994a). Meiosis reinitiation from the first prophase is dependent on the levels of intracellular Ca^{2+} and pH in oocytes of the bivalves *Mactra chinensis* and *Limaria hakodatensis*. *Dev. Biol.* **166,** 587–599.

Deguchi, R., and Osanai, K. (1994b). Repetitive intracellular Ca^{2+} increases at fertilization and the role of Ca^{2+} in meiosis reinitiation from the first metaphase in oocytes of marine bivalves. *Dev. Biol.* **163,** 162–174.

Deguchi, R., Shirakawa, H., Oda, S., Mohri, T., and Miyazaki, S. (2000). Spatiotemporal analysis of Ca^{2+} waves in relation to the sperm entry site and animal-vegetal axis during Ca^{2+} oscillations in fertilized mouse eggs. *Dev. Biol.* **218,** 299–313.

Digonnet, C., Aldon, D., Leduc, N., Dumas, C., and Rougier, M. (1997). First evidence of a calcium transient in flowering plants at fertilization. *Development* **124,** 2867–2874.

Ducibella, T. (1996). The cortical reaction and development of activation competence in mammalian oocytes. *Hum. Reprod. Update* **2,** 29–42.

Ducibella, T. (1998). Biochemical and cellular insights into the temporal window of normal fertilization. *Theriogenology* **49,** 53–65.

Ducibella, T., Rangarajan, S., and Anderson, E. (1988). The development of mouse oocyte cortical reaction competence is accompanied by major changes in cortical vesicles and not cortical granule depth. *Dev. Biol.* **130,** 789–792.

Dupont, G. (1998). Link between fertilization-induced Ca^{2+} oscillations and relief from metaphase II arrest in mammalian eggs: A model based on calmodulin-dependent kinase II activation. *Biophys. Chem.* **72,** 153–167.

Dupont, G., McGuinness, O. M., Johnson, M. H., Berridge, M. J., and Borgese, F. (1996). Phospholipase C in mouse oocytes: Characterization of β and γ isoforms and their possible involvement in sperm-induced Ca^{2+} spiking. *Biochem. J.* **316,** 583–591.

Eckberg, W. R., and Miller, A. L. (1995). Propagated and nonpropagated calcium transients during egg activation in the annelid, *Chaetopterus*. *Dev. Biol.* **172,** 654–664.

Eisen, A., and Reynolds, G. T. (1984). Calcium transients during early development in single starfish (*Asterias forbesi*) oocytes. *J. Cell Biol.* **99,** 1878–1882.

Eisen, A., and Reynolds, G. T. (1985). Source and sinks for the calcium released during fertilization of single sea urchin eggs. *J. Cell Biol.* **100,** 1522–1527.

Epel, D. (1989). Arousal of activity in sea urchin eggs at fertilization. *In* "The Cell Biology of Fertilization" (H. Schatten and G. Schatten, eds.), pp. 361–385. Academic Press, San Diego.

Evans, J. P., Foster, J. A., McAvey, B. A., Gerton, G. L., Kopf, G. S., and Schultz, R. M. (2000). Effects of perturbation of cell polarity on molecular markers of sperm-egg binding sites on mouse eggs. *Biol. Reprod.* **62,** 76–84.

Fissore, R. A., Dobrinsky, J. R., Balise, J. J., Duby, R. T., and Robl, J. M. (1992). Patterns of intracellular Ca^{2+} concentrations in fertilized bovine eggs. *Biol. Reprod.* **47,** 960–969.

Fissore, R. A., Gordo, A. C., and Wu, H. (1998). Activation of development in mammals: Is there a role for sperm cytosolic factor? *Theriogenology* **49,** 43–52.

Fissore, R. A., Longo, F. J., Anderson, E., Parys, J. B., and Ducibella, T. (1999). Differential distribution of inositol trisphosphate receptor isoforms in mouse oocytes. *Biol. Reprod.* **60,** 49–57.

Fissore, R. A., and Robl, J. M. (1993). Sperm, inositol trisphosphate, and thimerosal-induced intracellular Ca^{2+} elevations in rabbit eggs. *Dev. Biol.* **159,** 122–130.

Freeman, G., and Ridgway, E. B. (1993). The role of intracellular calcium and pH during fertilization and egg activation in the hydrozoan *Phialidium*. *Dev. Biol.* **156,** 176–190.

Fujiwara, T., Nakada, K., Shirakawa, H., and Miyazaki, S. (1993). Development of inositol trisphosphate-induced calcium release mechanism during maturation of hamster oocytes. *Dev. Biol.* **156,** 69–79.

Gardiner, D. M., and Grey, R. D. (1983). Membrane junctions in *Xenopus* eggs: Their distribution suggests a role in calcium regulation. *J. Cell Biol.* **96,** 1159–1163.

Gardner, R. L. (1999a). Polarity in early mammalian development. *Curr. Opin. Gene. Dev.* **9,** 417–421.

Gardner, R. L. (1999b). Scrambled or bisected mouse eggs and the basis of patterning in mammals. *Bioessays* **21,** 271–274.

Gilkey, J. C., Jaffe, L. F., Ridgway, E. B., and Reynolds, G. T. (1978). A free calcium wave traverses the activating egg of the medaka, *Oryzias latipes*. *J. Cell Biol.* **76,** 448–466.

Giusti, A. F., Carroll, D. J., Abassi, Y. A., and Foltz, K. R. (1999). Evidence that a starfish egg Src family tyrosine kinase associates with PLC-γ1 SH2 domains at fertilization. *Dev. Biol.* **208,** 189–199.

Goud, P. T., Goud, A. P., Van Oostveldt, P., and Dhont, M. (1999). Presence and dynamic redistribution of type I inositol 1,4,5-trisphosphate receptors in human oocytes and embryos during in-vitro maturation, fertilization and early cleavage divisions. *Mol. Hum. Reprod.* **5,** 441–451.

Goudeau, M., and Goudeau, H. (1996). External Mg^{2+} triggers oscillations and a subsequent sustained level of intracellular free Ca^{2+}, correlated with changes in membrane conductance in the oocyte of the prawn *Palaemon serratus*. *Dev. Biol.* **177,** 178–189.

Han, J., and Nuccitelli, R. (1990). Inositol 1,4,5-trisphosphate-induced calcium release in the organelle layers of the stratified, intact egg of *Xenopus laevis*. *J. Cell Biol.* **110,** 1103–1110.

He, C. L., Damiani, P., Parys, J. B., and Fissore, R. A. (1997). Calcium, calcium release receptors, and meiotic resumption in bovine oocytes. *Biol. Reprod.* **57,** 1245–1255.

Herbert, M., Gillespie, J. I., and Murdoch, A. P. (1997). Development of calcium signalling mechanisms during maturation of human oocytes. *Mol. Hum. Reprod.* **3,** 965–973.

Hirai, S., Nagahama, Y., Kishimoto, T., and Kanatani, H. (1981). Cytoplasmic maturity revealed by the structural changes in incorporated spermatozoon during the course of starfish oocyte maturation. *Dev. Growth Differ.* **23,** 465–478.

Hirose, K., Kadowaki, S., Tanabe, M., Takeshima, H., and Iino, M. (1999). Spatiotemporal dynamics of inositol 1,4,5-trisphosphate that underlies complex Ca^{2+} mobilization patterns. *Science* **284,** 1527–1530.

Jaconi, M., Pyle, J., Bortolon, R., Ou, J., and Clapham, D. (1997). Calcium release and influx colocalize to the endoplasmic reticulum. *Curr. Biol.* **7,** 599–602.

Jaffe, L. A., and Terasaki, M. (1993). Structural changes of the endoplasmic reticulum of sea urchin eggs during fertilization. *Dev. Biol.* **156,** 566–573.

Jaffe, L. A., and Terasaki, M. (1994). Structural changes in the endoplasmic reticulum of starfish oocytes during meiotic maturation and fertilization. *Dev. Biol.* **164,** 579–587.

Jaffe, L. F. (1985). The role of calcium explosions, waves, and pulses in activating eggs. *In* "Biology of Fertilization" (C. B. Metz and A. Monroy, eds.), Vol. 3, pp. 127–165. Academic Press, San Diego.

Jones, K. T. (1998). Ca^{2+} oscillations in the activation of the egg and development of the embryo in mammals. *Int. J. Dev. Biol.* **42,** 1–10.

Jones, K. T., Carroll, J., Merriman, J. A., Whittingham, D. G., and Kono. (1995). Repetitive sperm-induced Ca^{2+} transients in mouse oocytes are cell cycle dependent. *Development* **121,** 3259–3266.

Jones, K. T., Cruttwell, C., Parrington, J., and Swann, K. (1998a). A mammalian sperm cytosolic phospholipase C activity generates inositol trisphosphate and causes Ca^{2+} release in sea urchin egg homogenates. *FEBS Lett.* **437**, 297–300.

Jones, K. T., Soeller, C., and Cannell, M. B. (1998b). The passage of Ca^{2+} and fluorescent markers between the sperm and egg after fusion in the mouse. *Development* **125**, 4627–4635.

Kimura, Y., Yanagimachi, R., Kuretake, S., Bortkiewicz, H., Perry, A. C. F., and Yanagimachi, H. (1998). Analysis of mouse oocyte activation suggests the involvement of sperm perinuclear material. *Biol. Reprod.* **58**, 1407–1415.

Kline, D. (1988). Calcium-dependent events at fertilization of the frog egg: Injection of a calcium buffer blocks ion channel opening, exocytosis, and formation of pronuclei. *Dev. Biol.* **126**, 346–361.

Kline, D., and Kline, J. T. (1992a). Repetitive calcium transients and the role of calcium in exocytosis and cell cycle activation in the mouse egg. *Dev. Biol.* **149**, 80–89.

Kline, D., and Kline, J. T. (1992b). Thapsigargin activates a calcium influx pathway in the unfertilized mouse egg and suppresses repetitive calcium transients in the fertilized egg. *J. Biol. Chem.* **267**, 17624–17630.

Kline, D., Mehlmann, L. M., Fox, C., and Terasaki, M. (1999). The cortical endoplasmic reticulum (ER) of the mouse egg: Localization of ER clusters in relation to the generation of repetitive calcium waves. *Dev. Biol.* **215**, 431–442.

Kline, J. T., and Kline, D. (1994). Regulation of intracellular calcium in the mouse egg: Evidence for inositol trisphosphate-induced calcium release, but not calcium-induced calcium release. *Biol. Reprod.* **50**, 193–203.

Kline, D., and Stewart-Savage, J. (1994). The timing of cortical granule fusion, content dispersal, and endocytosis during fertilization of the hamster egg: An electrophysiological and histochemical study. *Dev. Biol.* **162**, 277–287.

Kume, S., Muto, A., Aruga, J., Nakagawa, T., Michikawa, T., Furuichi, T., Nakade, S., Okano, H., and Mikoshiba, K. (1993). The *Xenopus* IP_3 receptor: Structure, function, and localization in oocytes and eggs. *Cell* **73**, 555–570.

Kume, S., Yamamoto, A., Inoue, T., Muto, A., Okano, H., and Mikoshiba, K. (1997). Developmental expression of the inositol 1,4,5-trisphosphate receptor and structural changes in the endoplasmic reticulum during oogenesis and meiotic maturation of Xenopus laevis. *Dev. Biol.* **182**, 228–239.

Lawrence, Y., Ozil, J. P., and Swann, K. (1998). The effects of a Ca^{2+} chelator and heavy-metal-ion chelators upon Ca^{2+} oscillations and activation at fertilization in mouse eggs suggest a role for repetitive Ca^{2+} increases. *Biochem. J.* **335**, 335–342.

Lawrence, Y., Whitaker, M., and Swann, K. (1997). Sperm-egg fusion is the prelude to the initial Ca^{2+} increase at fertilization in the mouse. *Development* **124**, 233–241.

Lee, C., Ferguson, M., and Chen, L. B. (1989). Construction of the endoplasmic reticulum. *J. Cell Biol.* **109**, 2045–2055.

Lee, K. W., Webb, S. E., and Miller, A. L. (1999). A wave of free cytosolic calcium traverses zebrafish eggs on activation. *Dev. Biol.* **214**, 168–180.

Lee, S. J., and Shen, S. S. (1998). The calcium transient in sea urchin eggs during fertilization requires the production of inositol 1,4,5-trisphosphate. *Dev. Biol.* **193**, 195–208.

Lindsay, L. L., Hertzler, P. L., and Clark, W. J. (1992). Extracellular Mg^{2+} induces an intracellular Ca^{2+} wave during oocyte activation in the marine shrimp *Sicyonia ingentis*. *Dev. Biol.* **152**, 94–102.

Machaty, Z., Funahashi, H., Day, B. N., and Prather, R. S. (1997). Developmental changes in the intracellular Ca^{2+} release mechanisms in porcine oocytes. *Biol. Reprod.* **56**, 921–930.

Maro, B., Howlett, S. K., and Webb, M. (1985). Non-spindle microtubule organizing centers in metaphase II-arrested mouse oocytes. *J. Cell Biol.* **101**, 1665–1672.

McDougall, A., and Sardet, C. (1995). Function and characteristics of repetitive calcium waves associated with meiosis. *Current Biology* **5**, 318–328.

McGuinness, O. M., Moreton, R. B., Johnson, M. H., and Berridge, M. J. (1996). A direct measurement of increased divalent cation influx in fertilised mouse oocytes. *Development* **122,** 2199–2206.

McPherson, S. M., McPherson, P. S., Mathews, L., Campbell, K. P., and Longo, F. J. (1992). Cortical localization of a calcium release channel in sea urchin eggs. *J. Cell Biol.* **116,** 1111–1121.

Mehlmann, L. M., Carpenter, G., Rhee, S. G., and Jaffe, L. A. (1998). SH2 domain-mediated activation of phospholipase Cγ is not required to initiate Ca^{2+} release at fertilization of mouse eggs. *Dev. Biol.* **203,** 221–232.

Mehlmann, L. M., and Kline, D. (1994). Regulation of intracellular calcium in the mouse egg: Calcium release in response to sperm or inositol trisphosphate is enhanced after meiotic maturation. *Biol. Reprod.* **51,** 1088–1098.

Mehlmann, L. M., Mikoshiba, K., and Kline, D. (1996). Redistribution and increase in cortical inositol 1,4,5-trisphosphate receptors after meiotic maturation of the mouse oocyte. *Dev. Biol.* **180,** 489–498.

Mehlmann, L. M., Terasaki, M., Jaffe, L. A., and Kline, D. (1995). Reorganization of the endoplasmic reticulum during meiotic maturation of the mouse oocyte. *Dev. Biol.* **170,** 607–615.

Miyazaki, S. (1995a). Calcium signalling during mammalian fertilization. *Ciba Found. Symp.* **188,** 235–247.

Miyazaki, S. (1995b). Inositol trisphosphate receptor mediated spatiotemporal calcium signalling. *Curr. Opin. Cell Biol.* **7,** 190–196.

Miyazaki, S., Hashimoto, N., Yoshimoto, Y., Kishimoto, T., Igusa, and Hiramoto, Y. (1986). Temporal and spatial dynamics of the periodic increase in intracellular free calcium at fertilization of golden hamster eggs. *Dev. Biol.* **118,** 259–267.

Miyazaki, S., and Hirai, S. (1979). Fast polyspermy block and activation potential. Correlated changes during oocyte maturation of a starfish. *Dev. Biol.* **70,** 327–340.

Miyazaki, S., Shirakawa, H., Nakada, K., and Honda, Y. (1993). Essential role of the inositol 1,4,5-trisphosphate receptor/Ca^{2+} release channel in Ca^{2+} waves and Ca^{2+} oscillations at fertilization of mammalian eggs. *Dev. Biol.* **158,** 62–78.

Miyazaki, S., Yuzaki, M., Nakada, K., Shirakawa, H., Nakanishi, Nakade, S., and Mikoshiba, K. (1992). Block of Ca^{2+} wave and Ca^{2+} oscillation by antibody to the inositol 1,4,5-trisphosphate receptor in fertilized hamster eggs. *Science* **257,** 251–255.

Mohri, T., Ivonnet, P. I., and Chambers, E. L. (1995). Effect on sperm-induced activation current and increase of cytosolic Ca^{2+} by agents that modify the mobilization of [Ca^{2+}]$_i$. I. Heparin and pentosan polysulfate. *Dev. Biol.* **172,** 139–157.

Nuccitelli, R. (1991). How do sperm activate eggs? *Curr. Topics Dev. Biol.* **25,** 1–16.

Nuccitelli, R., Yim, D. L., and Smart, T. (1993). The sperm-induced Ca^{2+} wave following fertilization of the *Xenopus* egg requires the production of Ins(1, 4, 5)P$_3$. *Dev. Biol.* **158,** 200–212.

Oberdorf, J. A., Head, J. F., and Kaminer, B. (1986). Calcium uptake and release by isolated cortices and microsomes from the unfertilized egg of the sea urchin *Strongylocentrotus droebachiensis*. *J. Cell Biol.* **102,** 2205–2210.

Oda, S., Deguchi, R., Mohri, T., Shikano, T., Nakanishi, S., and Miyazaki, S. (1999). Spatiotemporal dynamics of the [Ca^{2+}]$_i$ rise induced by microinjection of sperm extract into mouse eggs: Preferential induction of a Ca^{2+} wave from the cortex mediated by the inositol 1,4,5-trisphosphate receptor. *Dev. Biol.* **209,** 172–185.

Ozil, J. P. (1990). The parthenogenetic development of rabbit oocytes after repetitive pulsatile electrical stimulation. *Development* **109,** 117–127.

Ozil, J. P. (1998). Role of calcium oscillations in mammalian egg activation: Experimental approach. *Biophys. Chem.* **72,** 141–152.

Parrington, J., Brind, S., De Smedt, H., Gangeswaran, R., Lai, F. A., Wojcikiewicz, R., and Carroll, J. (1998). Expression of inositol 1,4,5-trisphosphate receptors in mouse oocytes and early embryos: The type I isoform is upregulated in oocytes and downregulated after fertilization. *Dev. Biol.* **203,** 451–461.

Parrington, J., Jones, K. T., Lai, A., and Swann, K. (1999). The soluble sperm factor that causes Ca^{2+} release from sea-urchin (*Lytechinus pictus*) egg homogenates also triggers Ca^{2+} oscillations after injection into mouse eggs. *Biochem. J.* **341,** 1–4.

Parys, J. B., McPherson, S. M., Mathews, L., Campbell, K. P., and Longo, F. J. (1994). Presence of inositol 1,4,5-trisphosphate receptor, calreticulin, and calsequestrin in eggs of sea urchins and *Xenopus laevis*. *Dev. Biol.* **161,** 466–476.

Pey, R., Vial, C., Schatten, G., and Hafner, M. (1998). Increase of intracellular Ca^{2+} and relocation of E-cadherin during experimental decompaction of mouse embryos. *Proc. Natl. Acad. Sci. USA* **95,** 12977–12982.

Ridgway, E. B., Gilkey, J. C., and Jaffe, L. F. (1977). Free calcium increases explosively in activating medaka eggs. *Proc. Natl. Acad. Sci. USA* **74,** 623–627.

Sardet, C., Roegiers, F., Dumollard, R., Rouviere, C., and McDougall, A. (1998). Calcium waves and oscillations in eggs. *Biophys. Chem.* **72,** 131–140.

Schatten, H., Schatten, G., Mazia, D., Balczon, R., and Simerly, C. (1986). Behavior of centrosomes during fertilization and cell division in mouse oocytes and in sea urchin eggs. *Proc. Natl. Acad. Sci. USA* **83,** 105–109.

Schatten, G., Simerly, C., Schatten, H. (1985). Microtubule configurations during fertilization, mitosis, and early development in the mouse and the requirement for egg microtubule-mediated motility during mammalian fertilization. *Proc. Natl. Acad. Sci. USA* **82,** 4152–4156.

Schultz, R. M., and Kopf, G. S. (1995). Molecular basis of mammalian egg activation. *Curr. Topics Dev. Biol.* **30,** 21–62.

Shen, S. S. (1995). Mechanisms of calcium regulation in sea urchin eggs and their activities during fertilization. *Curr. Topics Dev. Biol.* **30,** 63–101.

Shen, S. S., and Buck, W. R. (1993). Sources of calcium in sea urchin eggs during the fertilization response. *Dev. Biol.* **157,** 157–169.

Shiraishi, K., Okada, A., Shirakawa, H., Nakanishi, S., Mikoshiba, K., and Miyazaki, S. (1995). Developmental changes in the distribution of the endoplasmic reticulum and inositol 1,4,5-trisphosphate receptors and the spatial pattern of Ca^{2+} release during maturation of hamster oocytes. *Dev. Biol.* **170,** 594–606.

Shirakawa, H., and Miyazaki, S. (1995). Evidence for inositol tetrakisphosphate-activated Ca^{2+} influx pathway refilling inositol trisphosphate-sensitive Ca^{2+} stores in hamster eggs. *Cell Calcium* **17,** 1–13.

Snow, P., Yim, D. L., Leibow, J. D., Saini, S., and Nuccitelli, R. (1996). Fertilization stimulates an increase in inositol trisphosphate and inositol lipid levels in *Xenopus* eggs. *Dev. Biol.* **180,** 108–118.

Sousa, M., Barros, A., and Tesarik, J. (1996). The role of ryanodine-sensitive Ca^{2+} stores in the Ca^{2+} oscillation machine of human oocytes. *Mol. Hum. Reprod.* **2,** 265–272.

Speksnijder, J. E., Corson, D. W., Sardet, C., and Jaffe, L. F. (1989). Free calcium pulses following fertilization in the ascidian egg. *Dev. Biol.* **135,** 182–190.

Speksnijder, J. E., Sardet, C., and Jaffe, L. F. (1990). Periodic calcium waves cross ascidian eggs after fertilization. *Dev. Biol.* **142,** 246–249.

Speksnijder, J. E., Terasaki, M., Hage, W. J., Jaffe, L. F., and Sardet, C. (1993). Polarity and reorganization of the endoplasmic reticulum during fertilization and ooplasmic segregation in the ascidian egg. *J. Cell Biol.* **120,** 1337–1346.

Stachecki, J. J., and Armant, D. R. (1996a). Regulation of blastocoele formation by intracellular calcium release is mediated through a phospholipase C-dependent pathway in mice. *Biol. Reprod.* **55,** 1292–1298.

Stachecki, J. J., and Armant, D. R. (1996b). Transient release of calcium from inositol 1,4,5-trisphosphate-specific stores regulates mouse preimplantation development. *Development* **122,** 2485–2496.

Stephano, J. L., and Gould, M. C. (1997). The intracellular calcium increase at fertilization in *Urechis caupo* oocytes: Activation without waves. *Dev. Biol.* **191,** 53–68.

6. Endoplasmic Reticulum in Mammalian Eggs

Stewart-Savage, J., and Grey, R. D. (1982). The temporal and spatial relationships between cortical contraction, sperm trail formation, and pronuclear migration in fertilized *Xenopus* eggs. *Roux's Arch.* **191,** 241–245.

Stith, B. J., Espinoza, R., Roberts, D., and Smart, T. (1994). Sperm increase inositol 1,4,5-trisphosphate mass in *Xenopus laevis* eggs preinjected with calcium buffers or heparin. *Dev. Biol.* **165,** 206–215.

Stith, B. J., Goalstone, M., Silva, S., and Jaynes, C. (1993). Inositol 1,4,5-trisphosphate mass changes from fertilization through first cleavage in *Xenopus laevis. Mol. Biol. Cell* **4,** 435–443.

Stricker, S. A. (1995). Time-lapse confocal imaging of calcium dynamics in starfish embryos. *Dev. Biol.* **170,** 496–518.

Stricker, S. A. (1996). Repetitive calcium waves induced by fertilization in the nemertean worm *Cerebratulus lacteus. Dev. Biol.* **176,** 243–263.

Stricker, S. A. (1999). Comparative biology of calcium signaling during fertilization and egg activation in animals. *Dev. Biol.* **211,** 157–176.

Stricker, S. A., Centonze, V. E., and Melendez, R. F. (1994). Calcium dynamics during starfish oocyte maturation and fertilization. *Dev. Biol.* **166,** 34–58.

Stricker, S. A., Centonze, V. E., Paddock, S. W., and Schatten, G. (1992). Confocal microscopy of fertilization-induced calcium dynamics in sea urchin eggs. *Dev. Biol.* **149,** 370–380.

Stricker, S. A., Silva, R., and Smythe, T. (1998). Calcium and endoplasmic reticulum dynamics during oocyte maturation and fertilization in the marine worm *Cerebratulus lacteus. Dev. Biol.* **203,** 305–322.

Subramanian, K., and Meyer, T. (1997). Calcium-induced restructuring of nuclear envelope and endoplasmic reticulum calcium stores. *Cell* **89,** 963–971.

Sun, F. Z., Hoyland, J., Huang, X., Mason, W., and Moor, R. M. (1992). A comparison of intracellular changes in porcine eggs after fertilization and electroactivation. *Development* **115,** 947–956.

Swann, K., and Ozil, J. P. (1994). Dynamics of the calcium signal that triggers mammalian egg activation. *Int. Rev. Cytol.* **152,** 183–222.

Swann, K., and Parrington, J. (1999). Mechanism of Ca^{2+} release at fertilization in mammals. *J. Exp. Zool.* **285,** 267–275.

Taylor, C. T., Lawrence, Y. M., Kingsland, C. R., Biljan, M. M., and Cuthbertson, K. S. (1993). Oscillations in intracellular free calcium induced by spermatozoa in human oocytes at fertilization. *Hum. Reprod.* **8,** 2174–2179.

Terasaki, M., Chen, L. B., and Fujiwara, K. (1986). Microtubules and the endoplasmic reticulum are highly interdependent structures. *J. Cell Biol.* **103,** 1557–1568.

Terasaki, M., Henson, J., Begg, D., Kaminer, B., and Sardet, C. (1991). Characterization of sea urchin egg endoplasmic reticulum in cortical preparations. *Dev. Biol.* **148,** 398–401.

Terasaki, M., and Jaffe, L. A. (1991). Organization of the sea urchin egg endoplasmic reticulum and its reorganization at fertilization. *J. Cell Biol.* **114,** 929–940.

Terasaki, M., and Jaffe, L. A. (1993). Imaging endoplasmic reticulum in living sea urchin eggs. *Methods Cell Biol.* **38,** 211–220.

Terasaki, M., Jaffe, L. A., Hunnicutt, G. R., and Hammer, J. A. (1996). Structural change of the endoplasmic reticulum during fertilization: Evidence for loss of membrane continuity using the green fluorescent protein. *Dev. Biol.* **179,** 320–328.

Terasaki, M., and Sardet, C. (1991). Demonstration of calcium uptake and release by sea urchin egg cortical endoplasmic reticulum. *J. Cell Biol.* **115,** 1031–1037.

Tesarik, J., and Sousa, M. (1994). Comparison of Ca^{2+} responses in human oocytes fertilized by subzonal insemination and by intracytoplasmic sperm injection. *Fertil. Steril.* **62,** 1197–1204.

Turner, P. R., Sheetz, M. P., and Jaffe, L. A. (1984). Fertilization increases the polyphosphoinositide content of sea urchin eggs. *Nature* **310,** 414–415.

Wang, S., Gebre-Medhin, S., Betsholtz, C., Stalberg, P., Zhou, Y., Larsson, C., Weber, G., Feinstein, R., Oberg, K., Gobl, A., and Skogseid, B. (1998). Targeted disruption of the mouse phospholipase C beta3 gene results in early embryonic lethality. *FEBS Lett.* **441,** 261–265.

Whitaker, M., and Swann, K. (1993). Lighting the fuse at fertilization. *Development* **117,** 1–12.

Williams, C. J., Mehlmann, L. M., Jaffe, L. A., Kopf, G. S., and Schultz, R. M. (1998). Evidence that Gq family G proteins do not function in mouse egg activation at fertilization. *Dev. Biol.* **198,** 116–127.

Wu, H., He, C., Jehn, B., Black, S. J., and Fissore, R. A. (1998). Partial characterization of the calcium-releasing activity of porcine sperm cytosolic extracts. *Dev. Biol.* **203,** 369–381.

Xu, Z., Abbott, A., Kopf, G. S., Schultz, R. M., and Ducibella, T. (1997). Spontaneous activation of ovulated mouse eggs: Time-dependent effects on M-phase exit, cortical granule exocytosis, maternal messenger ribonucleic acid recruitment, and inositol 1,4,5-trisphosphate sensitivity. *Biol. Reprod.* **57,** 743–750.

Xu, Z., Kopf, G. S., and Schultz, R. M. (1994). Involvement of inositol 1,4,5-trisphosphate-mediated Ca^{2+} release in early and late events of mouse egg activation. *Development* **120,** 1851–1859.

Yao, Y., Choi, J., and Parker, I. (1995). Quantal puffs of intracellular Ca^{2+} evoked by inositol trisphosphate in *Xenopus* oocytes. *J. Physiol.* **482,** 533–553.

Yue, C., White, K. L., Reed, W. A., and Bunch, T. D. (1995). The existence of inositol 1,4,5-trisphosphate and ryanodine receptors in mature bovine oocytes. *Development* **121,** 2645–2654.

Yue, C., White, K. L., Reed, W. A., and King, E. (1998). Localization and regulation of ryanodine receptor in bovine oocytes. *Biol. Reprod.* **58,** 608–614.

Zernicka-Goetz, M., Ciemerych, M. A., Kubiak, J. Z., Tarkowski, A. K., and Maro, B. (1995). Cytostatic factor inactivation is induced by a calcium-dependent mechanism present until the second cell cycle in fertilized but not in parthenogenetically activated mouse eggs. *J. Cell Sci.* **108,** 469–474.

Zucker, R. S., and Steinhardt, R. A. (1978). Prevention of the cortical reaction in fertilized sea urchin eggs by injection of calcium-chelating ligands. *Biochim. Biophys. Acta* **541,** 459–466.

7
Germ Plasm and Molecular Determinants of Germ Cell Fate

Douglas W. Houston and Mary Lou King
Department of Cell Biology and Anatomy
University of Miami School of Medicine
Miami, Florida 33101

I. Introduction
II. Germ Plasm
   A. History
   B. Germ Plasm in Early *Xenopus* Embryos
   C. Primordial Germ Cell Development and Migration
   D. Ultrastructure
III. Experimental Evidence for the Role of Germ Plasm in Primordial Germ Cell Formation
   A. *Drosophila*
   B. Amphibians
IV. Molecules Localized to the Germ Plasm
   A. Genes Involved in *Drosophila* Pole Cell Development
   B. P Granule Components in *C. elegans*
   C. *Xenopus* Germ Plasm-Localized RNAs
V. Summary and Future Directions
   A. Role of Germ Plasm in Primordial Germ Cell Formation
   B. Role of Localized Molecules in Germ Plasm
   References

One mechanism for the specification of cell types during embryonic development is the cytoplasmic localization of determinants in the egg into certain blastomeres. Primordial germ cell (PGC) development in many organisms is characterized by the inheritance of germ plasm, a cytologically distinct assembly of mitochondria and electron-dense germinal granules. This chapter reviews the structure of germ plasm and the experimental evidence for its importance in PGC specification in *Caenorhabditis elegans, Drosophila,* and *Xenopus*. It then compares and contrasts recent data on the identification of germ plasm components in these organisms. Many components are potentially RNA-binding proteins, implicating the regulation of RNA metabolism, transport, and translation as critical processes in PGC development. Germ plasm components also mediate transcriptional repression, regulate migration, and control mitotic divisions in PGCs. The chapter concludes with a discussion on the general roles of germ plasm components and how they might act to specify PGC fate. © 2000 Academic Press.

I. Introduction

In metazoan species, sexual reproduction is accomplished through syngamy, or fusion of two haploid gametes, the egg and sperm. The generation of haploid genomes via meiotic cell division and differentiation of the germ cells into functional gametes are critical processes in maintaining diversity and adaptability in organisms. As the germ cells are the agents of heredity, an understanding of the origins of the germ cells is fundamental to our understanding of reproduction.

Despite the vast diversity of organisms, the processes by which egg and sperm are formed are very similar. Germ cells arise outside the gonad early in development, in the form of primordial germ cells (PGCs). PGCs form within a variety of tissues, including the extraembryonic membranes of mammals, and begin migrating to the future genital ridges even before the gonads have formed. Upon arriving at the genital ridges, PGCs become ensheathed in the somatic gonad cells and begin the proliferation and differentiation that will produce functional gametes. PGCs are not sexually determined until they enter the gonad. Although the aspects of extragonadal PGC formation and migration are similar in many organisms, the specifics of how PGCs are formed differ greatly across the animal kingdom.

In general, PGCs can form as a result of interactions between tissues, i.e., epigenetically, or in a preformistic fashion, by inheritance of cytoplasmic determinants (reviewed in Nieuwkoop and Sutasurya, 1979, 1981). Although it is not clear what constrains some organisms to use a particular mode to specify PGCs, the hypothesis has been put forth that PGCs form in places that are the least influenced by adult body plan specification (Dixon, 1994). Epigenetic PGC origins were found in a variety of organisms in both vertebrate and invertebrate phyla. This mode of PGC formation is of interest because mammals, including mice and primates, appear to generate PGCs epigenetically.

Preformistic PGC specification is characterized by the asymmetric inheritance of maternal "germ cell determinants," localized in a cytologically distinct cytoplasm, or germ plasm. Although many invertebrate species, including insects, crustaceans, and nematodes, exhibit strongly preformistic modes of germ cell formation, only one vertebrate order, the anuran amphibians, are known to have germ plasm. Interestingly, urodele amphibians exhibit *epigenetic* PGC formation, prompting Nieuwkoop to propose a diphyletic origin of amphibia (Nieuwkoop and Sutasurya, 1979). Despite a wealth of both descriptive and experimental embryological evidence obtained over the past century, a clear picture of how germ plasm works is only beginning to emerge. This chapter places some of the recent advances in identifying the molecular components of germ plasm within the context of classical research on the structure and function of germ plasm. It focuses on important developments in *Xenopus*, *Drosophila*, and *Caenorhabditis elegans*. For more specific reviews on flies and worms, we refer the reader to Williamson and Lehmann (1996) and Seydoux and Strome (1999).

II. Germ Plasm

A. History

The idea of cytoplasmic germ cell determinants has a long history in embryology. Indeed, the role of cytoplasmic components in development was already appreciated in the late 1800s, when the functions of the nucleus and cytoplasm were first being distinguished. Classic work in the mollusc, *Dentalium* (Wilson, 1925), demonstrated that removal of the polar lobe during cleavage could block the formation of some larval structures. Additional work by Conklin (1905), in the ascidian *Styela*, demonstrated five cytoplasmic regions in the egg, each responsible for producing different tissues. Included among these regions is the conspicuous "yellow crescent" of the myoplasm, which localizes to the posterior of the egg and specifies tail muscle formation. These and other results not covered here, established early in the history of experimental embryology that elements in the cytoplasm were critical in governing cell fate during development.

Observations of uniquely staining granules and cytoplasmic inclusions in the primordial germ cells led to the suspicion that cytoplasmic localization was also important in germ cell formation. Several authors found deeply stained regions in the eggs and pole cells of dipteran insects and proposed roles for these regions in the segregation of pole cells. Kahle (1908) traced the fate of pole cells in *Miastor* by the presence of a darkly stained "polares Plasma." This appears to be the origin of the term pole plasm, which is still used to describe the abdomen and germ cell-determining region of insect eggs. Huettner (1921) found a similar pole plasm with distinct granules in *Drosophila*. Experimental evidence implicated these cytoplasmic determinants in PGC formation. Notably, Hegner (1911) ablated the pole plasm and pole cells from chrysomelid beetle (Coleoptera) eggs with a hot needle and found that the resulting embryos lacked germ cells. Additionally, more modern ablation experiments, described later, have helped answer the question of whether areas of cytoplasm are required for the formation of PGCs. However, the central question of how the germ plasm specifies PGC fate remains unanswered.

B. Germ Plasm in Early *Xenopus* Embryos

The phenomenon of cytoplasmic determination of germ cells seemed to be limited to invertebrates until Bounoure (1931, 1934) found a substance similar to pole plasm in the eggs of *Rana* and followed its presence into the PGCs of embryos. Bounoure referred to this substance as "cytoplasme germinal" or germ plasm. Nieuwkoop and Faber (1956) described a similar germ plasm, or "cytoplasmic inclusions," identical to Bounoure's "cytoplasme germinal" in the eggs and early embryos of *Xenopus laevis*. Blackler (1958) examined germ plasm in a variety of anuran amphibians to confirm and extend Bounoure's observations. He also determined, by histological methods, the presence of mitochondria and "pentosenucleic acids" (RNA) in the germ plasm.

Whitington and Dixon (1975) made the most detailed study of germ plasm in *Xenopus*. The germ plasm is found in many discrete islands in the vegetal hemisphere. These islands aggregate at the vegetal pole following fertilization and are then divided into separate blastomeres during cleavage (Figs. 1 and 2; see color plates). Germ plasm aggregation was found to depend on microtubules (Ressom and Dixon, 1988; Savage and Danilchik, 1993) and the activity of a kinesin-like protein, Xklp-1 (Robb *et al.*, 1996). Once organized into large patches, the germ plasm associates with the cleavage membranes and follows the ingressing cytoplasm into the interior of the embryo. By the blastula stage, germ plasm is normally found in three to seven cells near the floor of the blastocoel. Throughout early cleavage and the blastula stage, the germ plasm is unequally segregated into only one daughter cell, possibly via association with one pole of the spindle apparatus. During gastrulation, the germ plasm changes its position in the cell and comes to surround the nucleus. Subsequently, the germ plasm becomes equally distributed between daughter cells and a stable population of PGCs is established.

Aside from the resulting equal distribution of germ plasm, any other significance of perinuclear germ plasm positioning remains unknown. Whitington and Dixon (1975) thought it was indicative of the germ plasm becoming "active" in the determination of PGCs. Blackler (1958) also believed that PGCs were determined when the germ plasm became perinuclear. He reasoned that most other embryonic cells are also specified during gastrulation and observed that the germ plasm loses its cytochemical staining for RNA during this time.

C. Primordial Germ Cell Development and Migration

1. *Xenopus*

During gastrulation and neurulation in *Xenopus*, PGCs follow the normal morphogenetic movements of the endodermal mass and accumulate posteriorly in the endoderm (Whitington and Dixon, 1975; Figs. 1 and 2). These movements are most likely passive in nature. The germ plasm remains perinuclear during this time and can still be identified with basophilic stains. By the late tailbud stages (stages 30–35), the PGCs move laterally and then dorsally in the endoderm until they reach the dorsal crest of the endoderm at the early tadpole stage (stage 40). These migratory processes are believed to be the result of active rather than passive movement (Nieuwkoop and Sutasurya, 1979). At this stage, the germ plasm within PGCs is reported to lose its affinity for basophilic stains. However, PGCs can still be identified by morphological criteria such as lobed nuclei, the presence of small yolk platelets, and evidence of a distinct extracellular space

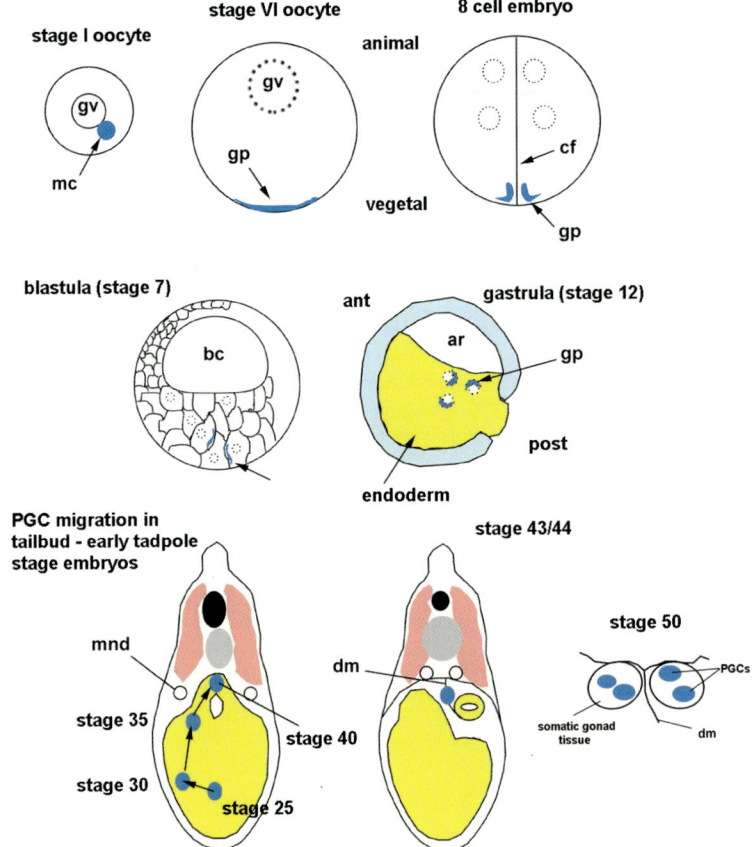

Chapter 7, Fig. 1 Life history of germ plasm and germ cells in *Xenopus*. Germ plasm (gp) initially accumulates in the mitochondrial cloud (mc) of stage I oocytes and is displaced to the vegetal cortex of stage VI oocytes. Following fertilization and during the cleavage stages (8-cell embryo), the germ plasm is concentrated at the vegetal pole and is internalized by ingressing cytoplasm and formation of the cleavage furrows (cf). In the blastula stage, the germ plasm remains associated with the plasma membranes of vegetal blastomeres. During the gastrula stage, the germ plasm becomes perinuclear and gastrulation movements position the PGCs in the posterior endoderm. The PGCs remain in this region until migration begins in the tailbud stages. PGCs first move laterally and then dorsally until they reach the dorsal crest of the endoderm by stage 40. As the dorsal mesentery (dm) forms, the PGCs leave the endoderm and migrate along the dorsal mesentery to the dorsal body wall. Once there, they move laterally and associate with somatic gonadal cell precursors and form the complete gonads (stage 50). n, nucleus; gv, germinal vesicle; bc, blastocoel; ar, archenteron; ant, anterior; post, posterior; mnd, mesonephric duct. Dark blue, germ plasm and PGCs; light blue, gastrula ectoderm and mesoderm; yellow, endoderm and gut; red, somites; black, neural tube; gray, notochord. Diagrams represent sections through oocytes and embryos.

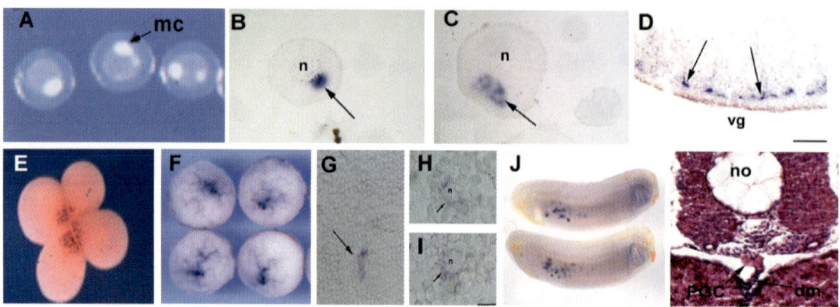

Chapter 7, Fig. 2 Germ cell development in *Xenopus*. (A) The mitochondrial cloud (mc) of stage I *Xenopus* oocytes. (B–J) Whole mount *in situ* hybridization of germ plasm-localized RNAs in oogenesis and embryogenesis. (B) Section showing expression of *Xdazl* RNA in the mitochondrial cloud of a stage I oocyte (arrow). (C) *Xdazl* in a stage II oocyte showing a wedge-like pattern (arrow). (D) Expression of *Xdazl* near the vegetal cortex of a stage VI oocyte (arrows). vg, vegetal pole. (E) Vegetal view of a 4-cell embryo stained for *Xcat2* RNA. (F) Vegetal views of stage 7 embryos stained for *DEADSouth* RNA. (G) Section of a stage 7 embryo showing *Xdazl* staining localized in the germ plasm (arrow). (H and I) Perinuclear *Xdazl* staining in stage 10 embryos (arrows). (J) *Xpat* staining of PGCs in tailbud stage embryos (stage 24). (K) Hematoxylin and eosin-stained section through a stage 44 embryo showing PGCs in the dorsal mesentery (dm). no, notochord. Scale bars: 62.5 µm.

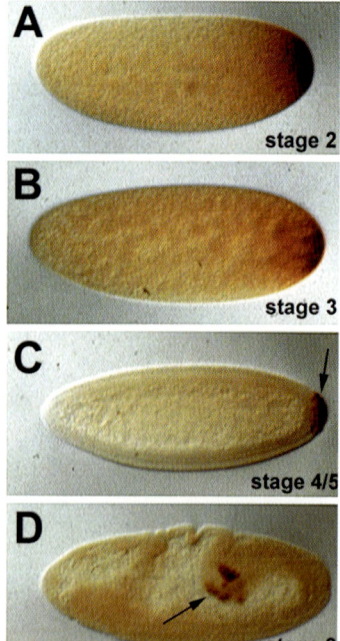

Chapter 7, Fig. 5 Expression of Nanos protein *Drosophila* embryos. (A and B) Stage 2 and 3 early embryos showing distribution of Nanos in the posterior region of the embryos. By the cellular blastoderm (C, stage 4/5), Nanos protein is incorporated into the pole cells (arrow). (D) Pole cells, containing Nanos protein, localized in the posterior midgut. These pictures were kindly contributed by Dr. Ruth Lehmann.

7. Germ Plasm Components

surrounding the PGCs (Nieuwkoop and Sutasurya, 1979). Subsequent migration of the PGCs to the gonads takes place via the dorsal mesentery, a thin strip of connective tissue that links the dorsal body wall to the gut. The mesentery forms from two sheets of splanchnic mesoderm surrounding the gut. As these sheets converge at the dorsal crest of the endoderm, PGCs exit the endoderm and incorporate into the mesentery (Whitington and Dixon, 1975). At this point (stage 43/44), the PGCs are easily recognizable in sections due their large size, lobed nuclei with prominent nucleoli, and yolk content. PGC numbers have increased to approximately 30 by this stage, as a result of two to three cell divisions (Dziadek and Dixon, 1977). From the mesentery, the PGCs then migrate laterally to the forming genital ridges (Wylie and Heasman, 1976), enter the gonads, and begin the process gametogenesis.

2. *Drosophila*

A similar process occurs in *Drosophila* (reviewed in Williamson and Lehmann, 1996). Pole cells form after 10 divisions of the syncitial nuclei of the embryo. Nuclei that enter the posterior pole plasm cellularize in advance of the rest of the syncitial blastoderm and incorporate the pole plasm. These pole cells undergo approximately two mitoses and attach to the presumptive midgut endoderm during gastrulation. Pole cell mitosis is then arrested until the PGCs enter the gonad. A process called germ band extension carries the pole cells, along with the midgut endoderm, dorsoanteriorly and deposits them inside the embryo. The pole cells then actively migrate posteriorly through the midgut pocket and through the posterior midgut tissue to the site of the developing gonadal mesoderm tissue. Definitive gonads form by what is termed gonad coalescence, in which the somatic gonad tissue surrounds the pole cells. Although there are many important differences between PGC migration in *Drosophila* and *Xenopus*, which will be discussed later, two important similarities stand out. First, the PGCs form early by association of germ plasm with nuclei. Second, PGCs migrate by combining passive attachment with active movement and by changing interactions with endodermal and mesodermal germ layers.

3. *C. elegans*

Through the use of direct observation and ablation experiments, the complete embryonic cell lineage of *C. elegans* has been established (Sulston *et al.*, 1983; Fig. 3). The germ cells are derived clonally from the P4 founder blastomere, which arises from four asymmetric cell divisions. Division of P4 yields the primordial germ cells, termed Z2 and Z3. These cells do not migrate as in other organisms. However, they must associate with and become surrounded by the somatic gonadal cells (Z1 and Z4) in order for germ cell proliferation to proceed

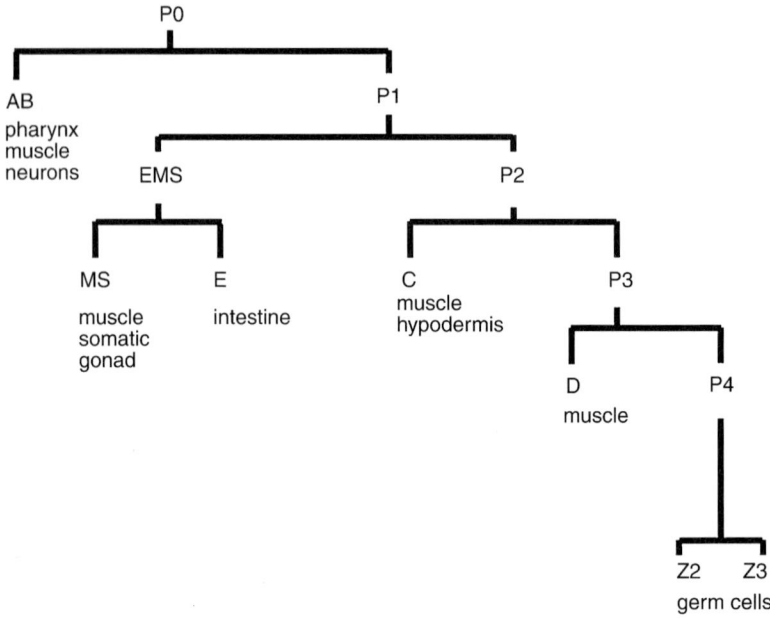

Fig. 3 Early embryonic cell lineage of *C. elegans*.

during subsequent larval stages. In the case of *C. elegans*, migration may not be necessary because the gonad develops adjacent to the PGCs, although some cell rearrangements may occur. Prior to entering the gonad, Z2 and Z3 appear to have extended contact with intestinal cells in a manner reminiscent of the association of PGCs with gut tissue in *Drosophila* and *Xenopus*. It is interesting to speculate that cell–cell contact with the endoderm is needed in some way to keep PGCs competent to enter the gonad, possibly through influencing the expression of cell surface proteins.

D. Ultrastructure

The fine structure of the germ plasm has been investigated in many organisms (for reviews, see Beams and Kessel, 1971; Nieuwkoop and Sutasurya, 1979, 1981). The main characteristics are quite similar across species but give few clues to the mechanism of germ plasm function. Germ plasm is enriched in two major components, mitochondria and unique granules or "germinal granules" (Fig. 4). In *Drosophila*, these are termed polar granules. The granules are seen by electron microscopy as electron-dense bodies with a granular or fibrillar subunit composition. They are often found in association with clusters of mito-

7. Germ Plasm Components

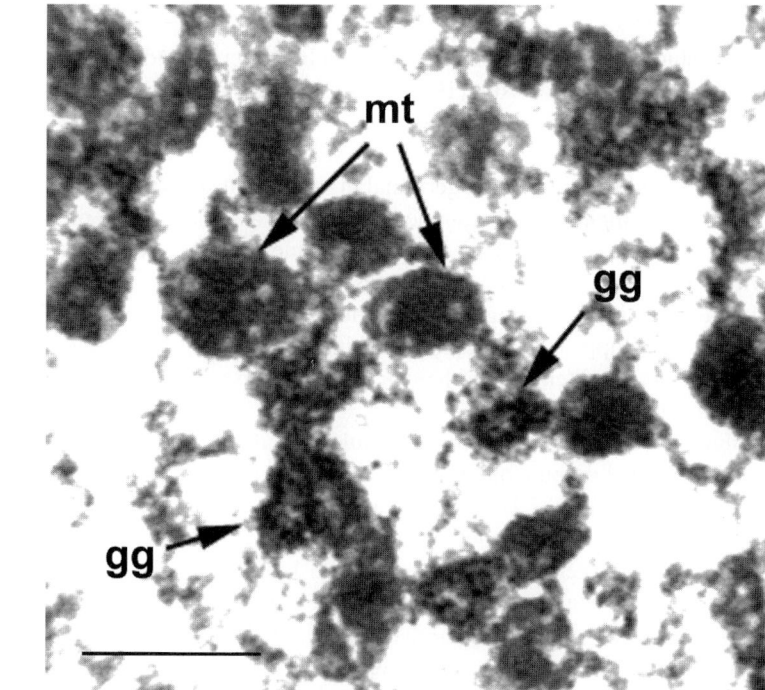

Fig. 4 Electron micrograph showing germ plasm in a stage I *Xenopus* oocyte. Note the numerous mitochondria (mt) and electron-dense germinal granule material (gg). Scale bar: 500 nm.

chondria or the nuclear envelope, depending on the stage of germ cell development, and are sometimes seen in transit through nuclear pores.

1. *Drosophila*

In *Drosophila*, polar granules appear in the pole plasm of oocytes during vitellogenesis as approximately 1-μm electron-dense bodies without a surrounding membrane (Mahowald, 1968). The bodies are composed of fibrils 15–20 nm thick. During fertilization and cleavage, the granules lose their association with mitochondria, become fragmented, and begin to associate with ribosomes. Following pole cell formation, the polar granules coalesce and lose their ribosomes. Interestingly, Mahowald (1971) found a dramatic decrease in the amount of cytochemically detectable RNA in the polar granules after pole cell formation. The protein content, indicated to be basic proteins, remains unchanged. At the end of germ band extension, the polar granules again fragment and become associated with the nuclear envelope. Additionally, the granules lose their "granular" appearance and are seen as a less compact mass of thin fibrils approximately 5 nm in diameter. These fibrils remain in association with the nuclear pores.

2. Amphibians

Subsequent studies in Anurans confirmed a similar structure and life history for amphibian germ plasm. Balinsky (1966) was the first to observe electron-dense granules in the germ plasm of amphibian eggs. In *Rana*, (Mahowald and Hennen, 1971; Williams and Smith, 1971) and *Xenopus* (Czolowska, 1972; Kalt, 1973; Ikenishi and Kotani, 1975), germinal granules of a similar size and structure to that of *Drosophila* polar granules were found associated with mitochondria in the germ plasm of unfertilized eggs. In some cases, a central lumen was also evident. The germinal granules increase in size during the blastula stage, possibly due to fusion or aggregation of granules. In a manner similar to *Drosophila*, the granules appear to undergo a transition to a "string-like" morphology during tailbud stages (Ikenishi and Kotani, 1975) and associate with the nuclear envelope during PGC migration.

Although the continuity of germinal granules was demonstrated in *Drosophila*, whether germinal granule material is present continuously in *Xenopus* is less clear. Williams and Smith (1971) found that germinal granules were formed during oocyte maturation from smaller, electron-dense areas in the vegetal pole. This result suggested that germinal granule material is present in mature oocytes. Looking at ovary tissue at earlier stages of oogenesis, Al-Mukhtar and Webb (1971) observed "nuage" material, ultrastructurally similar to germ plasm, associated with a perinuclear mitochondrial aggregate in gonadal PGCs and oogonia. This aggregate is a precursor to the mitochondrial cloud, a site of mitochondrial replication and a conspicuous structure present in previtellogenic oocytes (stage I) in *Xenopus* (Dumont, 1972; Billett and Adam, 1976). During stage II of oogenesis, this structure fragments and a large portion of it localizes to the future vegetal pole, the eventual site of the germ plasm.

Strong evidence for the mitochondrial cloud as the origin of germinal granule material was obtained by Heasman *et al.* (1984). They described an electron-dense, granulofibrillar material (GFM) in the mitochondrial cloud (Fig. 4). This material remained with the cloud during its fragmentation and became localized with aggregations of mitochondria in the subcortical vegetal cytoplasm. The amount of GFM appeared to increase in the previtellogenic stages, suggesting that the material was synthesized during this time. Heasman *et al.* (1984) also provide descriptive evidence that GFM is derived from nuage material within the mitochondrial cloud. Thus, the germinal granule material appears in germ cells of all types: in oogonia as perinuclear nuage, in previtellogenic oocytes in mitochondrial cloud GFM, and in mature oocytes, eggs, and embryos as germinal granules. These terms are largely based on descriptions and no functional consequence of the morphological changes in the germinal granule material is known.

3. *C. elegans*

Although the work of Boveri (1910) established that cytoplasmic factors were responsible for germ cell determination in nematodes, there were no descriptions

7. Germ Plasm Components

of a typical germ plasm. Ultrastructural analysis of the nematode *C. elegans* (Krieg *et al.*, 1978) identified "electron-light cytoplasmic areas" adjacent to nuclei of P lineage cells. Subsequent improvements in fixation methods led to the description of typical germinal granules in P lineage cells (Wolf *et al.*, 1983). Strome and Wood (1982, 1983) used immunofluorescent staining to trace the P granules throughout the germ cell life cycle. P granules are asymmetrically segregated into the germline precursor cells during the first four cleavages and associate with the nuclear envelope from P4 through germ cell proliferation and into the meiotic stages.

4. Concluding Remarks

From the results described earlier, it seems likely that the basic material of germinal granules is present continuously in germ cells and PGCs. The form of this material changes depending on the stage in development, as does its association with mitochondria and nuclei. Unfortunately, very little functional information about germ plasm can be gleaned from these studies. All of the changes in germinal granule shape and position could be merely the consequence of germ cell differentiation. However, knowledge of these morphological changes can provide clues as to when and where molecular alterations may be occurring, indicating critical times in germ cell life. Obviously, more detailed knowledge of the molecular composition of germinal granules is needed. Mahowald (1968) made perhaps the most prescient observation of the time; he suspected that the germinal granules contained translationally regulated maternal mRNAs required for germ cell specification.

III. Experimental Evidence for the Role of Germ Plasm in Primordial Germ Cell Formation

A. *Drosophila*

As mentioned previously, initial experiments in Coleoptera demonstrated the importance of germ plasm in the development of germ cells. Other experiments, initiated by Geigy (1931), showed that ultraviolet (UV) irradiation of the posterior pole, but not the anterior pole, of *Drosophila* embryos caused sterility and failure of pole cell formation (reviewed in Nieuwkoop and Sutasurya, 1981). The determinative nature of pole plasm was clearly demonstrated by Illmensee and Mahowald (1974). They transferred the posterior pole plasm from *Drosophila* early embryos to the anterior pole of host embryos of the same stage. Characteristic "pole cells" formed in this ectopic location and could produce functional germ cells when transplanted to the posterior pole of a genetically different host. Pole plasm also induced functional pole cells when transplanted to the midventral region of embryos (Illmensee and Mahowald, 1976). Interestingly, ectopic pole cells were able to reach the gonads and produce functional germ cells

(Illmensee and Mahowald, 1976). In the penultimate version of these experiments, which will be discussed later, ectopic expression of *oskar* RNA, a pole plasm component, at the anterior pole was able to organize polar granules and induce pole cell formation there (Ephrussi and Lehmann, 1992). Underwood *et al.* (1980) showed that transplanted *Drosophila* pole cells contributed only to germ cells, not to any of the primary germ layers. Thus the pole plasm appears to act as a determinant of germ cell fate. Although these types of experiments demonstrate that factors in the posterior cytoplasm are necessary and sufficient for pole cell formation, it cannot be discerned whether the polar granules themselves are critical.

B. Amphibians

Both ablation and overexpression experiments were also performed with anuran amphibian embryos to demonstrate the importance of the germ plasm. Bounoure (1937), who first described amphibian germ plasm in *Rana*, UV irradiated the vegetal poles of fertilized eggs. Although the irradiated animals were not completely sterile, germ cell numbers were reduced by approximately 90%. Smith (1966) performed an extensive analysis of UV irradiation effects on PGC formation in *Rana* embryos. He defined the UV-sensitive period as ranging from prior to first cleavage through second cleavage. Importantly, Smith found that injection of vegetal pole cytoplasm, but not animal pole cytoplasm, into irradiated embryos could rescue PGC formation. By varying the wavelength of UV light, Smith determined that 254 nm was most effective. This led to the speculation that the UV-sensitive target was DNA or RNA (Smith, 1966). Similar experiments were subsequently performed in *Xenopus* with comparable results (Ikenishi *et al.*, 1974; Züst and Dixon, 1975; Wakahara, 1977). Other ablation experiments, such as physical removal of vegetal pole cytoplasm, were also successful in causing sterility or partial sterility (Buehr and Blackler, 1970). These experiments indicate that as in *Drosophila*, *Xenopus* germ plasm is necessary for PGC formation. Other studies, however, suggest that anuran germ plasm may not be sufficient to form PGCs.

While germ plasm is capable of rescuing PGC formation following UV irradiation (Smith, 1966; Wakahara, 1977) or causing supernumerary PGCs (Wakahara, 1978) when injected into the vegetal pole, it appears to have no activity when injected ectopically. These observations support the notion that for germ plasm to function, it must be inherited by vegetal cells that are destined to lie in the posterior endoderm. Cleine and Dixon (1985) found that when *Xenopus* eggs were rotated either 90° or 180° after fertilization, the position of the germ plasm was altered relative to the cleavage planes. Thus the germ plasm came to lie in equatorial cells that were fated to lie in the anterior rather than posterior endoderm. The result of this shift was that fewer PGCs migrated to the genital ridges.

7. Germ Plasm Components

These experiments suggest that the influence of the endodermal environment on the action of germ plasm is greater in *Xenopus* than in other organisms.

Consistent with this idea, the behavior of PGCs in cell transplantation studies differs significantly between *Drosophila* and *Xenopus*. As described earlier, pole cells transplanted to the midventral region could migrate to the gonad and form germ cells (Illmensee and Mahowald, 1976). Similar studies in *Xenopus*, using either migratory PGCs (stage 45) or germ plasm containing blastomeres (stage 7) placed into a host embryo's blastocoel, found that these transplanted cells contributed to all three germ layers (Wylie *et al.*, 1985). In the same assay, vegetal blastomeres became restricted to the endoderm by the gastrula stage, suggesting that pluripotency is lost in nongerm plasm-containing cells but retained in PGCs (Wylie *et al.*, 1987). In support of this idea, *Rana* PGC nuclei were found capable of supporting full development in nuclear transplantation experiments, whereas endoderm nuclei could not (Smith, 1965). These studies indicate that in Anurans, germ plasm probably does not irreversibly restrict the developmental capacity of PGCs, as is the case in flies. Rather, a combination of germ plasm and the environment of the posterior endoderm may be needed for the correct differentiation and migration of frog PGCs.

Both ablation and transplantation experiments have been used to demonstrate the importance of germ plasm in germ cell formation. However, the germ plasm appears to act as a cell autonomous determinant in *Drosophila* pole cells, whereas in *Xenopus*, it has a more permissive role. Thus, frog germ plasm may only function properly when in the correct endodermal environment. As the germ plasm structure is very similar in the two species, the differences in action may be due to fundamental divergences in overall developmental mechanisms. Alternatively, differences in the types of localized molecular components could contribute to contrasting activities of germ plasm in the different organisms.

IV. Molecules Localized to the Germ Plasm

Important advances in the study of germ plasm have been made along several avenues of research. Genetic analysis of mutants in *Drosophila* and *C. elegans* identified a number of genes required for either the assembly or the function of germ plasm (reviewed in Wylie, 1999). Additional studies on localized antigens or RNAs in these organisms, as well as *Xenopus*, led to the identification of even more important germ plasm components (reviewed in Wylie, 1999). However, surprisingly few of these components are directly involved in either germ cell formation or development. This section summarizes what is known about localized molecules in germ plasm (Table I) and how these components function in early PGC development. Because many of these genes are involved in processes other than PGC formation, we have excluded or limited the discussion of those genes involved in germ plasm assembly, or those required

TABLE I Molecules Localized to the Germ Plasm

RNA/protein	Description/defects	References
Drosophilia		
Nanos	Zinc finger protein, required for pole cell migration, mitotic arrest, and repression of *Sxl* expression in pole cells	Kobayashi et al. (1996), Forbes and Lehmann (1998), Deshpande et al. (1999)
pgc-1	Noncoding RNA, required for pole cell migration	Nakamura et al. (1996)
gcl	Encodes nuclear pore protein, involved in pole cell formation	Jongens et al. (1992, 1994)
mtlrRNA	Rescues pole cell formation in UV-irradiated embryos	Kobayashi and Okada (1989), Iida and Kobayashi (1998)
Vasa	eIF4A homology, required for pole plasm assembly, involved in regulation of translation	Hay et al. (1990), Markussen et al. (1995), Rongo et al. (1995), Gavis et al. (1996), Carrera et al. (2000)
Tudor	Novel protein, required for polar plasm assembly	Bardsley et al. (1993)
C. elegans		
PIE-1	Zinc finger, required for transcriptional repression in early embryos	Mello et al. (1996), Seydoux et al. (1996), Seydoux and Dunn (1997)
MEX-1	Zinc finger, required for segregation of PIE-1 and P granules	Guedes and Priess (1997)
POS-1	Zinc finger, required for translation of APX-1	Tabara et al. (1998)
MEX-3	KH domain RNA-binding protein, involved in blastomere identity, mutation results in ectopic germ cells	Draper et al. (1996)
PGL-1	RGG-box protein, required for germ cell proliferation in larvae	Kawasaki et al. (1996)
GLH-1	Vasa-related DEAD-box helicase, needed for localization of PGL-1 to P granules	Gruidl et al. (1996), in Kawasaki et al. (1996)
Xenopus		
Xcat2	Encodes a Nanos-like zinc finger, protein expressed in germ plasm of gastrula, RNA is in germinal granules	Mosquera et al. (1993), Kloc et al. (1998)
mtlrRNA	Localized to germ plasm and germinal granules	Kobayashi et al. (1998)
Xlsirts	Contains interspersed repeated elements, not in germinal granules, noncoding and coding transcripts	Kloc et al. (1993, 1998)
Xwnt11	Encodes a Wnt protein, RNA not localized to germinal granules	Ku and Melton (1993), Kloc et al. (1998)
Xpat	Encodes a novel protein	Hudson and Woodland (1998)
DEADSouth	Encodes an eIF4A-like helicase	MacArthur et al., in press
Xdazl	Homologous to DAZ-related genes, protein expressed in the germ plasm, required for PGC differentiation	Houston et al. (1998), Houston and King (2000)

in the somatic cells for PGC migration or those involved in germ cell proliferation in the gonads.

A. Genes Involved in *Drosophila* Pole Cell Development

Numerous maternal effect mutations affecting pole plasm formation have been identified in *Drosophila* (reviewed in Rongo and Lehmann, 1996). Mutant embryos in this group lack pole cells and polar granules are also deficient in abdomen formation. Many of these genes, however, are required for the correct localization of *oskar* (*osk*) RNA, the critical component in pole plasm assembly. *osk* RNA and protein have a quantitative effect on the number of pole cells that form, suggesting that *osk* may directly control the amount of pole plasm that is made. Additionally, misexpression of *osk* RNA at the anterior pole causes the formation of ectopic polar granules and functional pole cells (Ephrussi and Lehmann, 1992). The proteins Vasa and Tudor are also required for the pole plasm assembly and recruitment of the so-called germ plasm effector genes, *nanos* (*nos*), *germ cell-less* (*gcl*), the 16S mitochondrial large rRNA (*mtlrRNA*) (reviewed in Williamson and Lehmann, 1996), and a nontranslated RNA, *polar granule component* (*pgc*, Nakamura *et al.*, 1996). Pole plasm thus serves two important functions in *Drosophila:* localization of the posterior determinant *nos* to the correct location and recruitment of factors for germ cell development.

Nanos, a zinc finger protein, is synthesized exclusively from RNA localized to the pole plasm and becomes incorporated into pole cells as they cellularize (Fig. 5; see color plate). This maternal protein remains expressed in the pole cells throughout embryogenesis. Pole cells deficient in Nanos form normally, but are unable to enter the gonads in transplantation experiments (Kobayashi *et al.*, 1996). Forbes and Lehmann (1998) produced embryos lacking maternal Nanos and found that *nos* was required for proper pole cell migration. Furthermore, Kobayashi *et al.* (1996) and Asaoka *et al.* (1998) demonstrated, using enhancer trap lines, that some promoters normally expressed in germ cells after gonad coalescence are activated earlier in pole cells lacking *nos* activity. Consistent with this result, the pair-rule genes *fushi-tarazu* (*ftz*) and *even-skipped* (*eve*) and the *Sex-lethal* (*Sxl*) gene were found ectopically expressed in pole cells lacking Nanos (Deshpande *et al.*, 1999). Interestingly, removal of *Sxl* could alleviate some of the migration and cell cycle defects in *nos* mutants, whereas ectopic expression of *Sxl* in wild-type pole cells mimicked these defects. These results suggest that one of the key functions of maternal Nanos is to repress the expression of genes that would otherwise interfere with the proper development of the pole cells.

Nanos, together with Pumilio, acts to repress *hunchback* translation during embryogenesis, allowing abdomen development. In studies on Pumilio, Asaoka-Taguchi *et al.* (1999) found that the Nos/Pum complex also represses translation of *cyclin B* RNA, thus preventing mitosis in migrating pole cells. Deshpande *et al.* (1999) also found that *nos* mutant pole cells fail to arrest mitosis following the cellular blastoderm stage, resulting in slightly more pole cells. Thus, two activities of Nos involve repression of translation; however, it is not known whether Nanos acts by this mechanism to suppress *Sxl* or by affecting *Sxl* transcription directly.

pgc was identified in a differential screen for pole cell-specific RNAs (Nakamura *et al.*, 1996). *In situ* hybridization at the ultrastructural level showed that *pgc* RNA was highly concentrated in polar granules. In embryos expressing an antisense *pgc* RNA construct, pole cells fail to populate the gonads in a manner similar to *nos* mutants. Because *pgc* is not predicted to encode a protein, it may play a role as a structural component of polar granules or as a regulatory RNA. The roles of *gcl* and *mtlrRNA* are also unclear, but they appear to function in pole cell formation or cellularization. *gcl* encodes a nuclear pore protein whose RNA localizes to the pole plasm (Jongens *et al.*, 1992, 1994). Overexpression of *gcl* at the posterior pole or inhibition with antisense RNA results in correspondingly more or fewer pole cells formed (Jongens *et al.*, 1992, 1994). Recent analysis of a *gcl* null mutation (Robertson *et al.*, 1999) indicated that Gcl was critical for pole cell formation, although embryos with very few pole cells were found in about 30% of the cases. Also, nuclear envelope-localization of Gcl protein was shown to be required for pole cell formation, but not for early steps in pole cell cellularization (pole bud formation). One reason for this discrepancy could be that the later events in pole cell formation require some communication between the nucleus and the forming pole cell membrane.

In the pole plasm, *mtlrRNA* localizes outside the mitochondria, in association with polar granules (Kobayashi *et al.*, 1993). Injection of *mtlrRNA* into the posterior pole of UV-irradiated embryos can restore pole cell formation, but not fertility (Kobayashi and Okada, 1989). This observation could indicate that *mtlrRNA* is one of the targets of irradiation or can substitute functionally for a bonafide target. To determine if *mtlrRNA* functions in pole cell formation, several groups looked for the appearance of *mtlrRNA* in ectopic polar granules induced by the anterior expression of *osk* (Ding *et al.*, 1994; Kobayashi *et al.*, 1995). These two studies obtained conflicting results. Subsequently, Iida and Kobayashi (1998) reduced the level of *mtlrRNA* using ribozymes and found that pole cells failed to form in a small percentage ($\sim$13%) of these embryos. However, the specificity of the ribozymes was not controlled for by the subsequent injection of *mtlrRNA* transcripts in rescue experiments. Thus the question of the requirement for *mtlrRNA* remains unresolved. It should be mentioned that the small ribosomal RNA, *mtsrRNA*, is also concentrated in the pole plasm (Ding *et al.*, 1994; Kashikawa *et al.*, 1999) and on polar granules (Kashikawa *et al.*, 1999). Mitochondrial RNAs may exit the mitochondria and form polar granule-specific translation machinery; however, the transit of these molecules from the mitochondria has not been demonstrated. The formal possibility that these RNAs are transcribed from mtDNA that has translocated to the nuclear genome has been ruled out.

Given the power of *Drosophila* genetics, it is interesting that few mutations affecting pole cell development have been isolated. Just like the case of *nos*, genes important for pole cell development may be critical for other processes as well, obscuring their roles in later development. Similarly, the functions in germ cell development of two maternal proteins localized to the pole plasm, Vasa and Tudor, have been difficult to establish. This is because maternal effect mutations in *vasa* and *tudor* disrupt *osk* localization, and hence polar granule assembly and pole cell formation (reviewed in Williamson and Lehmann, 1996). Vasa, an initiation factor-like RNA helicase, appears also to be involved in the translation of *osk* (Markussen *et al.*, 1995; Rongo *et al.*, 1995) and *nos* (Gavis *et al.*, 1996) within the pole plasm. The Vasa protein interacts with translation initiation dIF2 (Carrera *et al.*, 2000), demonstrating that Vasa is involved directly in regulating translation. Tudor is a novel protein with no known function. However, Tudor-like domains are found in many proteins that bind RNA (Ponting, 1997), suggesting that Tudor may have a role in RNA metabolism or transport. Nongenetic approaches, such as the use of dominant negative constructs, may be useful in determining the postpolar granule assembly functions of these proteins.

B. P Granule Components in *C. elegans*

Through a combination of genetic screens, homology cloning, and identification of P granule antigens, numerous genes have been isolated that affect *C. elegans* early germ cell development (reviewed in Wylie, 1999; Seydoux and Strome, 1999). Like *Drosophila*, many of these genes have poorly understood functions and have multiple roles in either somatic development or P granule segregation, in addition to roles in germ cell development. Note that all of the following genes can potentially encode RNA-binding proteins.

Several genes encoding zinc finger proteins play critical roles in early germ cell development in *C. elegans* (reviewed in Seydoux and Strome, 1999). PIE-1, MEX-1, and POS-1 all contain unusual CCCH zinc fingers, but are otherwise unrelated. Except for PIE-1, which is also nuclear, these proteins are cytoplasmic and associate with P granules in early embryogenesis (Fig. 6). In a surprising series of findings, P lineage cells were found to lack new transcription during early cleavage stages, and this transcriptional repression was shown to be dependent on the presence of the PIE-1 protein (Mello *et al.*, 1996; Seydoux *et al.*, 1996; Seydoux and Dunn, 1997). Sequence elements in PIE-1 resemble the C-terminal domain (CTD) of RNA polymerase II. Thus, PIE-1 mediated repression may act by targeting factors that interact with the CTD (Batchelder *et al.*, 1999). Transcriptional repression is thought to protect P lineage cells from somatic differentiation. This idea fits well with the "enclave hypothesis" for germ plasm (see Section V,B). Interestingly, *Drosophila* pole cells also appear to undergo a period of transcriptional repression (Zalokar, 1976; Seydoux and Dunn, 1997; Van Doren

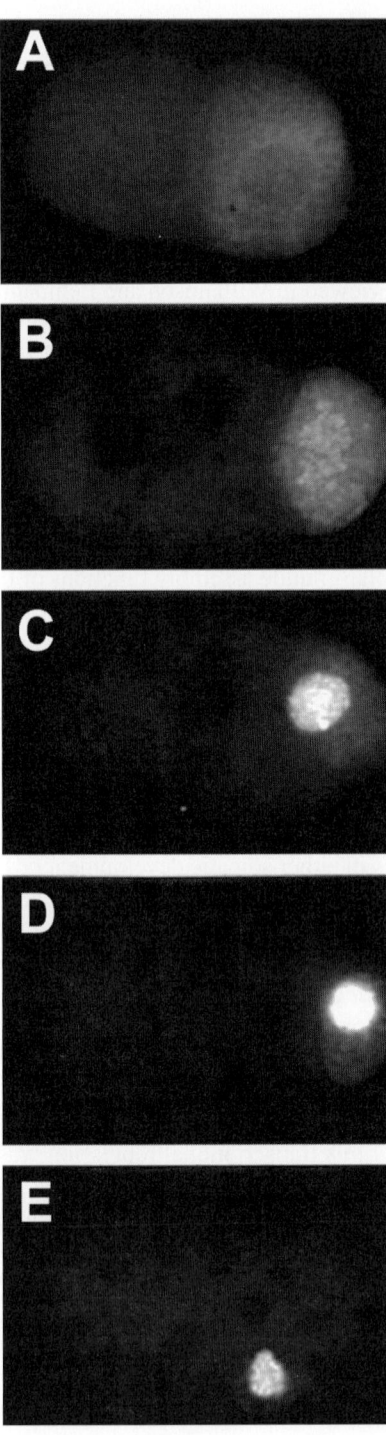

7. Germ Plasm Components

et al., 1998), suggesting that preventing new transcription might be a general mechanism for specifying germ cell fate.

MEX-1 is required for the proper localization of PIE-1 to P lineage cells (Guedes and Priess, 1997). Transcriptional repression is not disrupted in *mex-1* mutants; indeed, ectopic PIE-1 protein seems to repress transcription abnormally in somatic blastomeres. However, germ cells still fail to form, suggesting a distinct role for MEX-1 in germ cell formation. POS-1 was shown to be required for the translation of APX-1, a signaling protein required for early patterning. *pos-1* mutants also do not segregate P granules properly and may initiate transcription in P lineage cells earlier than normal, similar to *nos* mutations in *Drosophila* (Tabara *et al.*, 1998). Mutations in *pie-1*, *mex-1*, and *pos-1* are all maternal effect lethal, indicating that somatic fates are affected as well as the germ cells.

The *mex-3* gene was also identified as a maternal effect lethal mutation and encodes a KH domain-type RNA-binding protein (Draper *et al.*, 1996). Although MEX-3 is expressed in all blastomeres, the protein associates with P granules during early cleavages, suggesting a role in germ cell development. *mex-3* embryos appear to contain ectopic germ cells; however, the significance of this observation is unknown and has not been discussed in the literature.

Another set of genes identified in *C. elegans* encode proteins that localize exclusively to the P granules and are involved only in germ cell development. The *pgl-1* gene was identified in a screen for mutants that failed to stain for a P granule-specific monoclonal antibody (Kawasaki *et al.*, 1998). The PGL-1 protein contains an RGG-box RNA-binding motif and is expressed on P granules at all stages of development. Analysis of *pgl-1* mutants revealed a temperature-dependent defect in late larval germ cell proliferation. Early germ cell development and P granule integrity were not affected.

The *germline helicase* genes (*glh-1* and *glh-2*) were identified by homology to *Drosophila vasa*. The GLH-1 and GLH-2 proteins are found on the P granules and are possible RNA helicases containing DEAD-box helicase motifs and several CCHC zinc fingers (Gruidl *et al.*, 1996). They lack, however, an RGG-box found in Vasa. Unexpectedly, *glh-1* mutants show the identical germ cell proliferation phenotype as *pgl-1* mutants (unpublished, cited in Kawasaki, 1998). Furthermore, PGL-1 fails to localize to P granules in *glh-1* mutants, suggesting that GLH-1 is required for the localization of PGL-1 and that the sterile phenotype is a consequence of mislocalization of PGL-1 (unpublished, cited in Kawasaki,

---

Fig. 6 Expression of PIE-1 during early *C. elegans* development. (A) Two-cell embryo showing PIE-1 in the cytoplasm and nucleus of the P1 blastomere. (B) Early 4-cell embryo showing PIE-1 in the nucleus and cytoplasm of the P2 blastomere. By late 4-cell stage (C), PIE-1 staining it primarily nuclear. Note that this nuclear localization coincides with the beginning of transcriptional repression in the P lineage. PIE-1 remains nuclear in P3 at the 12-cell stage (D) and in P4 at the 24-cell stage. These pictures were kindly contributed by Dr. James R. Priess.

1998). It is possible that GLH-1 may play a similar role to Vasa in the localization and translation of germinal granule components.

Several homologues of *Drosophila nanos* have been identified in *C. elegans* (Subramaniam and Seydoux, 1999). *nos-2* RNA is localized to the P granules and is required for incorporation of PGCs into the somatic gonad. This function of *nos-2* is very similar to that of *Drosophila nos*. Also, like *Drosophila nos*, *nos-2* requires the activity of *pumilio* homologues (Subramaniam and Seydoux, 1999), suggesting that the mechanisms for controlling gonad entry are conserved between *Drosophila* and *C. elegans*.

In summary, studies of P granule components in *C. elegans* have identified a number of proteins with roles in germ cell determination. It is striking that all of these are potentially RNA-binding proteins, strongly suggesting that the regulation of RNA metabolism or translation is critical for the formation of primordial germ cells. Evidence to date points to roles for these components in translational control and transcriptional repression. The P granules are also known to contain maternal, polyadenylated RNAs (Seydoux and Fire, 1994), but aside from *nos-2*, these have yet to be characterized.

C. *Xenopus* Germ Plasm-Localized RNAs

In contrast to the case in *C. elegans*, a number of RNAs that localize to the germ plasm in *Xenopus* have been identified, although only few localized proteins are known (reviewed in Wylie, 1999; King *et al.*, 1999). Most germ plasm-localized RNAs were cloned on the basis of differential screening for vegetal pole or vegetal cortex RNAs. Only recently have the roles of these localized RNAs in PGC specification been examined in functional studies.

The first germ plasm-localized RNA discovered was *Xcat2* (Mosquera *et al.*, 1993), and it encodes a protein with Nanos-like zinc finger homology. This RNA was isolated from a library enriched in cytoskeleton-associated RNAs. The localization pattern of *Xcat2* RNA during oogenesis and early embryogenesis corresponds to that of the germ plasm (Fig. 2E). *Xcat2* first localizes to the mitochondrial cloud of stage I oocytes, as does the germ plasm, and is transported to the vegetal pole with the fragmenting cloud. The pattern of germ plasm RNA localization differs from that described for RNAs involved in early patterning such as *Vg1* (Rebatliati *et al.*, 1985) or *VegT* (Zhang and King, 1996). These latter RNAs localize late in oogenesis and require microtubules for efficient localization, whereas germ plasm RNAs are localized early by a microtubule-independent mechanism (reviewed in King *et al.*, 1999). In embryos, the maternal *Xcat2* RNA persists until the neurula stage (MacArthur *et al.*, 1999). The Xcat2 protein is detected in gastrula stage embryos, after the germ plasm has become perinuclear, suggesting that the maternal RNA is translation-

7. Germ Plasm Components

ally repressed during oogenesis (MacArthur *et al.*, 1999). It has been shown by *in situ* hybridization at the ultrastructural level that *Xcat2* RNA is concentrated in the germinal granule material of both stage I and stage IV oocytes (Kloc *et al.*, 1998). Given the conservation of function among *nos* homologs in *Drosophila* and *C. elegans*, it will be interesting to determine whether *Xcat2* has a similar role in *Xenopus*.

Xdazl encodes an RNA-binding protein homologous to the human Deleted in Azoospermia (DAZ) family of meiosis and gametogenesis regulators (Houston *et al.*, 1998). This family of proteins is involved in many processes related to germ cell development, including meiotic entry (*Drosophila* Boule; Eberhart *et al.*, 1996) and gamete differentiation (human DAZ/DAZL and mouse Dazl, see references in Houston *et al.*, 1998). Xdazl was shown to be required for early PGC differentiation (Houston and King, 2000). This is the first functional evidence for a role in *Xenopus* PGC development for a germ plasm-specific RNA. Embryos derived from oocytes that were depleted of maternal *Xdazl* RNA lack or are deficient in PGCs. This deficiency results from a failure of the PGCs to migrate out of the ventral endoderm. The approach taken in these studies was the antisense depletion and host transfer method of Heasman *et al.* (1991), which should also prove effective in identifying roles for other germ plasm RNAs in PGC development. Although it is not known exactly how Xdazl acts at the molecular level, it is likely that the regulation of translation is involved. Heterologous expression of the *Xdazl* cDNA in the testis of *boule* mutant flies results in a rescue of the meiotic entry phenotype (Houston *et al.*, 1998). Subsequently, Maines and Wasserman (1999) showed that Boule controls meiotic entry by regulating the translation of Twine, a Cdc25-like phosphatase. Because Xdazl can substitute for Boule, Xdazl must also regulate the translation of Twine. Whether or not Xdazl is also involved in the regulation of translation of proteins involved in PGC development remains to be established.

The other germ plasm-localized RNAs all show more or less the same localization pattern in oocytes as *Xcat2* and *Xdazl* (Fig. 2). *Xlsirts* are a group of both noncoding and coding RNAs that contain approximately 80 nucleotide tandemly repeated elements (Kloc *et al.*, 1993). The repeated units, ranging from 3 to 13 in number, are flanked by unrelated unique sequences. These sequences also appear unrelated to *Drosophila pgc-1* RNA, another noncoding RNA found in germ plasm. Unlike *pgc-1* and *Xcat2*, *Xlsirts* do not localize to the germinal granules, but are found in the germ plasm cytoplasm (Kloc *et al.*, 1998). The role of *Xlsirts* is unknown, although their absence from the germinal granules seems to rule out a possible structural role. *Xwnt11* encodes a secreted protein related to the Wnt family of signaling molecules (Ku and Melton, 1993). Like *Xlsirts*, *Xwnt11* localizes to the germ plasm but not within the germinal granules (Kloc *et al.*, 1998). *Xpat* encodes a novel protein (Hudson and Woodland, 1998). This maternal RNA is unusual in its expression pattern because it persists in PGCs until the early tadpole stage (stage 40) and thus serves as a very useful PGC-specific marker. *DEADSouth* is a DEAD-box helicase related to eIF4A (MacArthur *et al.*, 2000) and may function in PGC-specific translation initiation. Preliminary experiments suggest that unlike *Xcat2*, *DEADSouth* is not localized to germinal granules (M. L. King, unpublished observations). Similar to *Drosophila*, *Xenopus* germ plasm also contains *mtlrRNA* localized to the germinal granules (Kobayashi *et al.*, 1998); however, it is not yet known whether this RNA has any function in germ plasm.

V. Summary and Future Directions

A. Role of Germ Plasm in Primordial Germ Cell Formation

In general, there are several possible ways in which the inheritance of localized determinants could specify cell fate. In the extreme case, the determinants would be actual terminal differentiation products. Although this situation is unlikely, myoplasm of the ascidian *Styela* contains the mRNA for muscle-specific actin (Tomlinson *et al.*, 1987), so a role for terminal differentiation products cannot be completely excluded. Second, the localized determinants could contain "master regulators" that control a cascade of gene expression to control cell identity. The closest examples of this mechanism are *Drosophila bicoid* mRNA, which localizes to the anterior pole and specifies anterior development (Berleth *et al.*, 1988), and the *VegT* mRNA, a vegetal pole-localized RNA in *Xenopus* that is required for the formation of endoderm (Zhang *et al.*, 1998). The third alternative is that the determinants are intermediates between master genes and terminal differentiation. This situation seems to fit the germ plasm because of the numerous RNA-binding proteins; germ plasm could control gene expression at the posttranscriptional level. Indeed, analysis of *nanos* and *pos-1* mutants implicate regulation of translation as an important mechanism in PGC formation. Additionally, Vasa functions in translational regulation and Xdazl is likely to do so as well. Several *Drosophila* pole plasm RNAs, *cyclin B* and *gcl* (reviewed in Williamson and Lehmann, 1996), and *Xenopus* germ plasm RNAs, *Xcat2* (MacArthur *et al.*, 1999) and *Xdazl* (Houston and King, 2000), are under translational repression. It is possible that translational repression in the germ plasm serves to limit the expression of specific proteins until the PGC lineage can be established.

B. Role of Localized Molecules in Germ Plasm

Aside from a similar overall structure, there are several interesting common features among the germ plasm components in *Drosophila*, *C. elegans*, and *Xeno-*

7. Germ Plasm Components

pus. First, most of the proteins appear to be RNA-binding proteins. Second, *vasa*, *glh-1*, and *DEADSouth* all encode DEAD-box helicases with homology to eIF4A. Among these, Vasa and GLH-1 appear to be required for germinal granule formation, suggesting that helicases may function in regulating the localization of germinal granule components. These helicases could act in the packaging of components or by regulating the translation of structural elements. Third, germ plasm in both *Drosophila* and *Xenopus* contains both noncoding RNAs and mitochondrial RNAs. These RNAs could play a structural role in the germinal granules, in the case of *pgc-1*, or could be used as regulatory RNAs to sequester RNA-binding proteins. Mitochondrial RNAs are thought to mediate PGC-specific protein synthesis in the germinal granules. Another interesting commonality, the association of germ plasm with the nuclear envelope, suggests a role for germ plasm in the PGC-specific regulation of nuclear transport. For example, the *Drosophila* protein Gcl is a nuclear pore component (Jongens *et al.*, 1994). Also, P granules have been shown to associate with nuclear pore material in adult germ cells and retain this material in the cytoplasm after their detachment from the nuclear envelope (Pitt *et al.*, 2000). Whether this interesting observation is also true for *Xenopus* and *Drosophila* remains to be determined. Finally, the existence of *nanos* homologues in *C. elegans* and *Xenopus* suggests that the Nos/Pum RNA-binding complex is a critical component of germ plasm in both invertebrates and vertebrates. The identification of a *pumilio* homologue in *Xenopus* will be important in establishing the generality of *nos* function in germ plasm.

The predominant view in the literature about the function of germ plasm is that it acts as an "enclave" to "protect" PGCs from becoming somatic cells (Blackler, 1958; Smith, 1965; Dixon, 1994). Major questions to be resolved are how molecules localized to the germ plasm might accomplish this protection, and if protection alone is sufficient for PGC specification. Repression of transcription in the PGCs might be one mechanism by which these cells are protected from entering somatic lineages. Transcriptional repression is known to occur in *Drosophila* and *C. elegans* (Seydoux and Dunn, 1997). In *C. elegans*, PIE-1-mediated repression is thought to act globally, probably at the level of RNA pol II activation (Batchelder *et al.*, 1999). In *Drosophila*, however, a similar factor has not yet been identified; *nanos* mutants show ectopic transcription of only a few genes (Deshpande *et al.*, 1999) rather than a general alleviation of transcription quiescence (Seydoux and Dunn, 1997).

In *Xenopus*, it is currently not known whether transcriptional repression occurs in PGCs. The conservation of such a mechanism from invertebrates to vertebrates would indicate a general mode of action for germ plasm. *Xenopus* does differ from *Drosophila* with respect to the developmental potency of PGCs, suggesting that other mechanisms are at work. Pole cells are determined while frog PGCs retain pluripotency (Wylie *et al.*, 1985) and require positioning in the posterior endoderm. In this regard, *Xenopus* PGC development is more similar to the mouse, an animal without germ plasm.

Mouse PGCs develop from proximal epiblast cells, which are already pluripotent prior to gastrulation (reviewed in Pesce *et al.*, 1998). In this case, position is of prime importance; PGC precursors transplanted to a different location in the epiblast develop as somatic cells, whereas distal epiblast cells transplanted proximally can become PGCs (Tam and Zhou, 1996). A population of proximal epiblast cells, mostly fated to become extraembryonic mesoderm, are induced to form PGCs by signals from the extraembryonic trophectoderm, including the cytokine Bmp4 (Lawson *et al.*, 1999). In this case, the inducers are acting on a population of pluripotent cells to specify PGC fate. A similar situation may occur in *Xenopus*, with the posterior endoderm supplying the signals and the germ plasm conferring pluripotency to a small population of cells.

Understanding the role of germ plasm will undoubtedly require the identification of additional germ plasm components. Furthermore, it will be important to determine how these components interact with each other and how these elements in the cytoplasm ultimately affect cell fate.

Acknowledgments

We thank Dr. Ruth Lehmann for contributing pictures and Dr. James R. Priess for contributing figures and for communicating results prior to publication. This work was supported by a grant to M.L.K. from the National Institutes of Health and from the Human Frontier Science Program.

References

Al-Mukhtar, K. A. K., and Webb, A. C. (1971). An ultrastructural study of primordial germ cells, oogonia and early oocytes in *Xenopus laevis*. *J. Embryol. Exp. Morphol.* **26**, 195–217.

Asaoka, M., Sano, H., Obara, Y., and Kobayashi, S. (1998). Maternal Nanos regulates zygotic gene expression in germline progenitors of *Drosophila melanogaster*. *Mech. Dev.* **78**, 153–158.

Asaoka-Taguchi, M., Yamada, M., Nakamura, A., Hanyu, K., and Kobayashi, S. (1999). Maternal Pumilio acts together with Nanos in germline development in *Drosophila* embryos. *Nature Cell Biol.* **7**, 431–437.

Balinsky, B. I. (1966). Changes in the ultrastructure of amphibian eggs following fertilization. *Acta Embryol. Morphol. Exp.* **9**, 132–154.

Bardsley, A., McDonald, K., and Boswell, R. E. (1993). Distribution of tudor protein in the *Drosophila* embryo suggests separation of functions based on site of localization. *Development* **119**, 207–219.

Batchelder, C., Dunn, M. A., Choy, B., Suh, Y., Cassie, C., Shim, E. Y., Shin, T. H., Mello, C., Seydoux, G., and Blackwell, T. K. (1999). Transcriptional repression by *Caenorhabditis elegans* germ-line protein PIE-1. *Genes Dev.* **13**, 202–212.

Beams, H., and Kessel, R. (1974). The problem of germ cell determinants. *Int. Rev. Cytol.* **39**, 413–479.

Berleth, T., Burri, M., Thoma, G., Bopp, D., Richstein, S., Frigerio, G., Noll, M., and Nüsslein-Volhard, C. (1988). The role of localization of *bicoid* RNA in organizing the anterior pattern of the *Drosophila* embryo. *EMBO J.* **7**, 1749–1756.

7. Germ Plasm Components

Billet, F. S., and Adam, E. (1976). The structure of the mitochondrial cloud of *Xenopus laevis* oocytes. *J. Embryol. Exp. Morphol.* **33,** 697–710.

Blackler, A. W. (1958). Contribution to the study of germ-cells in the Anura. *J. Embryol. Exp. Morphol.* **6,** 491–503.

Bounoure, L. (1931). Sur l'existence d'un *déterminant germinal* dans l'oeuf indivis de la Grenouille rousse. *C. R. Acad. Sci. Paris* **193,** 402.

Bounoure, L. (1934). Recherches sur la lignée germinale chez la grenouille rousse aux premiers stades du développment. *Ann. Sci. Nat.* **17,** 67–248.

Bounoure, L. (1937). Le sort de la lignée germinale chez la Grenouille rousse après l'action des rayons ultra-violets sur le pôle inférieur de l'oeuf. *C. R. Acad. Sci. Paris* **204,** 1837.

Boveri, T. (1910). Die Potenzen der *Ascaris-Blastomeren*. *Festschr. f. R. Hertwig* **3,** 131.

Buehr, M. L., and Blackler, A. W. (1970). Sterility and partial sterility in the South African clawed toad following the pricking of the egg. *J. Embryol. Exp. Morphol.* **23,** 375–384.

Carrera, P., Johnstone, O., Nakamura, A., Casanova, J., Jäckle, H., and Lasko, P. (2000). VASA mediates translation through interaction with a *Drosophila* yIF2 homolog. *Mol. Cell* **5,** 181–187.

Cleine, J. H., and Dixon, K. E. (1985). The effect of egg rotation on the differentiation of primordial germ cells in *Xenopus laevis*. *J. Embryol. Exp. Morphol.* **90,** 79–99.

Conklin, E. G. (1905). Mosaic development in ascidian eggs. *J. Exp. Zool.* **2,** 145–223.

Czolowska, R. (1972). The fine structure of the "germinal cytoplasm" in the egg of *Xenopus laevis*. *Wilhelm Roux Arch. EntwMech. Org.* **169,** 335–344.

Deshpande, G., Calhoun, G., Yanowitz, J. L., and Schedl, P. D. (1999). Novel functions of *nanos* in downregulating mitosis and transcription during the development of the *Drosophila* germline. *Cell* **99,** 271–281.

Ding, D., Whittaker, K. L., and Lipshitz, H. D. (1994). Mitochondrially encoded 16S large ribosomal RNA is concentrated in the posterior polar plasm of early *Drosophila* embryos but is not required for pole cell specification. *Dev. Biol.* **163,** 503–15.

Dixon, K. E. (1994). Evolutionary aspects of primordial germ cell formation. *Ciba Found. Symp.* **182,** 92–120.

Draper, B. W., Mello, C. C., Bowerman, B., Hardin, J., and Priess, J. R. (1996). MEX-3 is a KH-domain protein that regulates blastomere identity in early *C. elegans* embryos. *Cell* **87,** 205–216.

Dumont, J. N. (1972). Oogenesis in *Xenopus laevis* (Daudin). I. Stages of oocyte development in laboratory maintained animals. *J. Morphol.* **136,** 153–180.

Dziadek, M., and Dixon, K. E. (1977). An autoradiographic analysis of nucleic acid synthesis in the presumptive primordial germ cells of *Xenopus laevis*. *J. Embryol. Exp. Morphol.* **37,** 13–31.

Eberhart, C. G., Maines, J. Z., and Wasserman, S. A. (1996). Meiotic cell cycle requirement for a fly homolog of human *Deleted in Azoospermia*. *Nature* **381,** 783–785.

Ephrussi, A., and Lehmann, R. (1992). Induction of germ cell formation by *oskar*. *Nature* **358,** 387–392.

Forbes, A., and Lehmann, R. (1998). Nanos and Pumilio have critical roles in the development and function of *Drosophila* germline stem cells. *Development* **125,** 679–690.

Gavis, E. R., Lunsford, L., Bergsten, S. E., and Lehmann, R. (1996). A conserved 90 nucleotide element mediates translational repression of *nanos* RNA. *Development* **122,** 2791–2800.

Geigy, R. (1931). Action de l'ultra-violet sur le pôle germinal dans l'oeuf de *Drosophila melanogaster* (castration et mutabilité). *Rev. Suisse Zool.* **38,** 187–288.

Gruidl, M. E., Smith, P. A., Kuznicki, K. A., McCrone, J. S., Kirchner, J., Roussell, D. L., Strome, S., and Bennett, K. L. (1996). Multiple potential germ-line helicases are components of the germ-line-specific P granules of *Caenorhabditis elegans*. *Proc. Natl. Acad. Sci. USA* **93,** 13837–13842.

Guedes, S., and Priess, J. R. (1997). The *C. elegans* MEX-1 protein is present in germline blastomeres and is a P granule component. *Development* **124,** 731–739.

Hay, B., Jan, L. Y., and Jan, Y. N. (1990). Localization of vasa, a component of *Drosophila* polar granules, in maternal-effect mutants that alter embryonic anteroposterior polarity. *Development* **109**, 425–433.

Heasman, J., Quarmby, J., and Wylie, C. C. (1984). The mitochondrial cloud of *Xenopus* oocytes: The source of the germinal granule material. *Dev. Biol.* **105**, 458–469.

Heasman, J., Holwill, S., and Wylie, C. C. (1991). Fertilization of cultured *Xenopus* oocytes and use in studies of maternally inherited molecules. *In* "Methods in Cell Biology" (B. K. Kay and H. B. Peng, eds.), pp. 685–695. Academic Press, New York.

Hegner, R. W. (1914). Studies on germ cells. I. The history of the germ cells in insects with special reference to the Keimbahn-determinants. II. The origin and significance of the Keimbahn-determinants in animals. *J. Morphol.* **25**, 375–509.

Holwill, S., Heasman, J., Crawley, C. R., and Wylie, C. C. (1987). Axis and germline deficiencies caused by u.v. irradiation of *Xenopus* oocytes cultured in vitro. *Development* **100**, 735–743.

Houston, D. W., and King, M. L. (2000). A critical role for *Xdazl*, a germ plasm-localized RNA, in the differentiation of primordial germ cells in *Xenopus*. *Development* **127**, 447–456.

Houston, D. W., Zhang, J., Maines, J. Z., Wasserman, S. A., and King, M. L. (1998). A *Xenopus* *DAZ*-like gene encodes an RNA component of germ plasm and is a functional homologue of *Drosophila boule*. *Development* **125**, 171–180.

Hudson, C., and Woodland, H. (1998). *Xpat*, a gene expressed specifically in germ plasm and primordial germ cells of *Xenopus laevis*. *Mech. Dev.* **73**, 159–168.

Huettner, A. P. (1921). The origin of the germ cells in *Drosophila melanogaster*. *J. Morphol.* **37**, 385–423.

Iida, T., and Kobayashi, S. (1998). Essential role of mitochondrially encoded large rRNA for germ-line formation in *Drosophila* embryos. *Proc. Natl. Acad. Sci. USA* **95**, 11274–78.

Ikenishi, K., and Kotani, M. (1975). Ultrastructure of the "germinal plasm" in *Xenopus* embryos after cleavage. *Dev. Growth Differ.* **17**, 101–110.

Ikenishi, K., Kotani, M., and Tanabe, K. (1974). Ultrastructural changes associated with UV irradiation in the "germinal plasm" of *Xenopus laevis*. *Dev. Biol.* **36**, 155–168.

Illmensee, K., and Mahowald, A. P. (1974). Transplantation of posterior polar plasm in *Drosophila*. Induction of germ cells at the anterior pole of the egg. *Proc. Natl. Acad. Sci. USA* **71**, 1016–1020.

Illmensee, K., and Mahowald, A. P. (1976). The autonomous function of germ plasm in a somatic region of the *Drosophila* egg. *Exp. Cell. Res.* **97**, 127–40.

Jongens, T., Ackerman, L., Swedlow, J., Jan, L., and Jan, Y. (1994). *Germ cell-less* encodes a cell type-specific nuclear pore associated protein and functions early in the germ-cell specification pathway of *Drosophila*. *Genes Dev.* **8**, 2123–2136.

Jongens, T. A., Hay, B., Jan, L. Y., and Jan, Y. N. (1992). The germ cell-less gene product: A posteriorly localized component necessary for germ cell development in *Drosophila*. *Cell* **70**, 569–584.

Kahle, W. (1908). Die Paedogenese der Cecidomyiden. *Zoologica* **21**, 1.

Kalt, M. R. (1973). Ultrastructural observations on the germ line of *Xenopus laevis*. *Z. Zellforsch. Mikrosk. Anat.* **138**, 41–62.

Kashikawa, M., Amikura, R., Nakamura, A., and Kobayashi, S. (1999). Mitochondrial small ribosomal RNA is present on polar granules in early cleavage embryos of *Drosophila melanogaster*. *Dev. Growth Differ.* **41**, 495–502.

Kawasaki, I., Shim, Y. H., Kirchner, J., Kaminker, J., Wood, W. B., and Strome, S. (1998). PGL-1, a predicted RNA-binding component of germ granules, is essential for fertility in *C. elegans*. *Cell* **94**, 635–645.

King, M. L., Zhou, Y., and Bubunenko, M. (1999). Polarizing genetic information in the egg: RNA localization in the frog oocyte. *BioEssays* **21**, 546–557.

Kloc, M., Larabell, C., Chan, A. P.-Y., and Etkin, L. D. (1998). Contribution of METRO pathway localized molecules to the organization of the germ cell lineage. *Mech. Dev.* **75**, 81–93.

7. Germ Plasm Components

Kloc, M., Spohr, G., and Etkin, L. D. (1993). Translocation of repetitive RNA sequences with the germ plasm in *Xenopus* oocytes. *Science* **262,** 1712–1714.

Kobayashi, S., Amikura, R., and Mukai, M. (1998). Localization of mitochondrial large ribosomal RNA in germ plasm of *Xenopus* embryos. *Curr. Biol.* **8,** 1117–1120.

Kobayashi, S., Amikura, R., Nakamura, A., Saito, H., and Okada, M. (1995). Mislocalizaton of *oskar* product in the anterior pole results in ectopic localization of mitochondrial large ribosomal RNA in *Drosophila* embryos. *Dev. Biol.* **169,** 384–386.

Kobayashi, S., Amikura, R., and Okada, M. (1993). Presence of mitochondrial large ribosomal RNA outside mitochondria in germ plasm of *Drosophila melanogaster*. *Science* **260,** 1521–1524.

Kobayashi, S., and Okada, M. (1989). Restoration of pole-cell-forming ability to u.v.-irradiated *Drosophila* embryos by injection of *mitochondrial lrRNA*. *Development* **107,** 733–742.

Kobayashi, S., Yamada, M., Asaoka, M., and Kitamura, T. (1996). Essential role of the posterior morphogen *nanos* for germline development in *Drosophila*. *Nature* **380,** 708–711.

Krieg, C., Cole, T., Deppe, U. Schierenberg, E., Schmitt, D., Yoder, B., and von Ehrenstein, G. (1978). The cellular anatomy of embryos of the nematode *Caenorhabditis elegans*: Analysis and reconstruction of serial section electron micrographs. *Dev. Biol.* **65,** 193–215.

Ku, M., and Melton, D. A. (1993). *Xwnt-11*: A maternally expressed *Xenopus* wnt gene. *Development* **119,** 1161–1173.

Lawson, K. A., Dunn, N. R., Roelen, B. A. J., Zeinstra, L. M., Davis, A. M., Wright, C. V. E., Korving, J. P. W. F. M., and Hogan, B. L. (1999). *Bmp4* is required for the generation of primordial germ cells in the mouse embryo. *Genes Dev.* **13,** 424–436.

MacArthur, H., Bubunenko, M., Houston, D. W., and King, M. L. (1999). *Xcat2* RNA is a translationally sequestered germplasm component in *Xenopus*. *Mech. Dev.* **84,** 75–88.

MacArthur, H., Houston, D., Bubunenko, M., Mosquera, L., and King, M. L. (2000). *DeadSouth* is a germ plasm specific DEAD-Box RNA helicase in *Xenopus* related to eIF4A. *Mech. Dev.,* in press.

Mahowald, A. P. (1968). Polar granules of *Drosophila*. II. Ultrastructural changes during early embryogenesis. *J. Exp. Zool.* **167,** 237–261.

Mahowald, A. P. (1971). Polar granules of *Drosophila*. IV. Cytochemical studies showing loss of RNA from polar granules during early stages of embryogenesis. *J. Exp. Zool.* **176,** 345–52.

Mahowald, A. P., and Hennen, S. (1971). Ultrastructure of the "germ plasm" in eggs and embryos of *Rana pipiens*. *Dev. Biol.* **24,** 37–53.

Maines, J. Z., and Wasserman, S. A. (1999). Post-transcriptional regulation of the meiotic Cdc25 protein Twine by the Dazl orthologue Boule. *Nat. Cell Biol.* **1,** 171–174.

Markussen, F.-H., Michon, A. M., Breitwieser, W., and Ephrussi, A. (1995). Translational control of *oskar* generates Short OSK, the isoform that induces pole plasm assembly. *Development* **121,** 3723–32.

Mello, C. C., Schubert, C., Draper, B., Zhang, W., Lobel, R., and Priess, J. R. (1996). The PIE-1 protein and germline specification in *C. elegans* embryos. *Nature* **382,** 710–712.

Mosquera, L., Forristall, C., Zhou, Y., and King, M. L. (1993). An mRNA localized to the vegetal cortex of *Xenopus* oocytes encodes a protein with a *nanos*-like zinc finger domain. *Development* **117,** 377–386.

Nakamura, A., Amikura, R., Mukai, M., Kobayashi, S., and Lasko, P. F. (1996). Requirement for a noncoding RNA in *Drosophila* polar granules for germ cell establishment. *Science* **274,** 2075–2079.

Nieuwkoop, P. D., and Faber, J. (1956). "Normal Table of *Xenopus laevis* (Daudin)," 1st Ed. North-Holland, Amsterdam.

Nieuwkoop, P. D., and Sutasurya, L. (1979). "Primordial Germ Cells in the Chordates." Cambridge Univ. Press, Cambridge, UK.

Nieuwkoop, P. D., and Sutasurya, L. (1981). "Primordial Germ Cells in the Invertebrates." Cambridge Univ. Press, Cambridge, UK.

Pesce, M., Gross, M. K., and Schöler, H. R. (1998). In line with our ancestors: *Oct-4* and the mammalian germ. *BioEssays* **20,** 722–732.
Pitt, J. N., Schisa, J. A., and Priess, J. R. (2000). P granules in the germ cells of *C. elegans* adults are associated with clusters of nuclear pores and contain RNA. *Dev. Biol.* **219,** 315–333.
Ponting, C. P. (1997). Tudor domains in proteins that interact with RNA. *Trends Biochem. Sci.* **22,** 51–2.
Rebatliati, M. R., Weeks, D. L., Harvey, R. P., and Melton, D. A. (1985). Identification and cloning of localized maternal RNAs from *Xenopus* eggs. *Cell* **42,** 769–777.
Ressom, R. E., and Dixon, K. E. (1988). Relocation and reorganization of germ plasm in *Xenopus* embryos after fertilization. *Development* **103,** 507–518.
Robb, D. L., Heasman, J., Raats, J., and Wylie, C. (1996). A kinesin-like protein is required for germ plasm aggregation in *Xenopus*. *Cell* **87,** 823–831.
Robertson, S. E., Dockendorff, T. C., Leatherman, J. L., Faulkner, D. L., and Jongens, T. A. (1999). *germ cell-less* is required only during the establishment of the germ cell lineage of *Drosophila* and has activities which are dependent and independent of its localization to the nuclear envelope. *Dev. Biol.* **215,** 288–297.
Rongo, C., Gavis, E. R., and Lehmann, R. (1995). Localization of *oskar* RNA regulates oskar translation and requires oskar protein. *Development* **121,** 2737–46.
Rongo, C., and Lehmann, R. (1996). Regulated synthesis, transport and assembly of the *Drosophila* germ plasm. *Trends Genet.* **12,** 102–109.
Savage, R. M., and Danilchik, M. V. (1993). Dynamics of germ plasm localizaiton and its inhibition by ultraviolet irradiation in early cleavage *Xenopus* embryos. *Dev. Biol.* **157,** 371–382.
Seydoux, G., and Dunn, M. A. (1997). Transcriptionally repressed germ cells lack a subpopulation of phosphorylated RNA polymerase II in early embryos of *Caenorhabditis elegans* and *Drosophila melanogaster*. *Development* **124,** 2191–2202.
Seydoux, G., and Fire, A. (1994). Soma-germline asymmetry in the distributions of embryonic RNAs in *Caenorhabditis elegans*. *Development* **120,** 2823–2834.
Seydoux, G., Mello, C. C., Pettitt, J., Wood, W. B., Priess, J. R., and Fire, A. (1996). Repression of gene expression in the embryonic germ lineage of *C. elegans*. *Nature* **382,** 713–716.
Seydoux, G., and Strome, S. (1999). Launching the germline in *Caenorhabditis elegans*: Regulation of gene expression in early germ cells. *Development* **126,** 3275–3283.
Smith, L. D. (1965). Transplantation of the nuclei of primordial germ cells into enucleated eggs of *Rana pipiens*. *Proc. Natl. Acad. Sci. USA* **54,** 101–107.
Smith, L. D. (1966). The role of a "germinal plasm" in the formation of primordial germ cells in *Rana pipiens*. *Dev. Biol.* **14,** 330–347.
Strome, S., and Wood, W. B. (1982). Immunofluorescence visualization of germ-line-specific cytoplasmic granules in embryos, larvae and adults of *Caenorhabditis elegans*. *Proc. Natl. Acad. Sci. USA* **79,** 1558–1562.
Strome, S., and Wood, W. B. (1983). Generation of asymmetry and segregation of germ-line granules in early *C. elegans* embryos. *Cell* **35,** 15–25.
Subrammaniam, K., and Seydoux, G. (1999). *nos-1* and *nos-2*, two genes related to *Drosophila nanos*, regulate primordial germ cell development and survival in *Caenorhabditis elegans*. *Development* **126,** 4861–4871.
Sulston, J. E., Schierenberg, E., White, J. G., and Thomson, J. N. (1983). The embryonic cell lineage of the nematode *Caenorhabditis elegans*. *Dev. Biol.* **100,** 64–119.
Tabara, H., Hill, R. J., Mello, C. C., Priess, J. R., and Kohara, Y. (1998). *pos-1* encodes a cytoplasmic zinc finger protein essential for germline specification in *C. elegans*. *Development* **126,** 1–11.
Tam, P. P. L., and Zhou, S. (1996). The allocation of epiblast cells to ectodermal and germ-line lineages is influenced by the position of the cells in gastrulating mouse embryos. *Dev. Biol.* **178,** 124–32.
Tomlinson, C. R., Bates, W. R., and Jeffrey, W. R. (1987). Development of a muscle actin specified by maternal and zygotic mRNA in ascidian embryos. *Dev. Biol.* **123,** 470–482.

Underwood, E. M., Caulton, J. H., Allis, C. D., and Mahowald, A. P. (1980). Developmental fate of pole cells in *Drosophila melanogaster*. *Dev. Biol.* **77,** 303–314.

Van Doren, M., Williamson, A. L., and Lehmann, R. (1998). Regulation of zygotic gene transcription in *Drosophila* primordial germ cells. *Curr. Biol.* **8,** 243–246.

Wakahara, M. (1977). Partial characterization of "primordial germ cell-forming activity" localized in vegetal pole cytoplasm in anuran eggs. *J. Embryol. Exp. Morphol.* **39,** 221–233.

Wakahara, M. (1978). Induction of supernumerary primordial germ cells by injecting vegetal pole cytoplasm into *Xenopus* eggs. *J. Exp. Zool.* **203,** 159–164.

Whitington, P. McD., and Dixon, K. E. (1975). Quantitative studies of germ plasm and germ cells during early embryogenesis of *Xenopus laevis*. *J. Embryol. Exp. Morphol.* **33,** 57–74.

Williams, M. A., and Smith, L. D. (1971). Ultrastructure of the "germinal plasm" during maturation and early cleavage in *Rana pipiens*. *Dev. Biol.* **25,** 568–580.

Williamson, A., and Lehmann, R. (1996). Germ cell development in *Drosophila*. *Annu. Rev. Cell Dev. Biol.* **12,** 365–91.

Wilson, E. B. (1925). "The Cell in Development and Heredity," 3rd Ed. MacMillan Company, New York.

Wolf, N., Priess, J., and Hirsh, D. (1983). Segregation of germline granules in early embryos of *Caenorhabditis elegans*: An electron microscope analysis. *J. Embryol. Exp. Morphol.* **73,** 297–306.

Wylie, C. (1999). Germ cells. *Cell* **96,** 165–174.

Wylie, C. C., and Heasman, J. (1976). The formation of the gonadal ridge in *Xenopus laevis*. I. A light and transmission electron microscope study. *J. Embryol. Exp. Morphol.* **35,** 125–138.

Wylie, C. C., Heasman, J., Snape, A., O'Driscoll, M., and Holwill, S. (1985). Primordial germ cells of *Xenopus laevis* are not irreversibly determined early in development. *Dev. Biol.* **112,** 66–72.

Wylie, C. C., Snape, A., Heasman, J., and Smith, J. C. (1987). Vegetal pole cells and their commitment to form endoderm in *X. laevis*. *Dev. Biol.* **119,** 496–502.

Zalokar, M. (1976). Autoradiographic study of protein and RNA formation during early development of *Drosophila* eggs. *Dev. Biol.* **49,** 425–437.

Zhang, J., Houston, D. W., King, M. L., Payne, C., Wylie, C., and Heasman, J. (1998). The role of maternal VegT in establishing the primary germ layers in *Xenopus* embryos. *Cell* **94,** 515–524.

Zhang, J., and King, M. L. (1996). *Xenopus VegT* RNA is localized to the vegetal cortex during oogenesis and encodes a novel T-box transcription factor involved in mesoderm patterning. *Development* **122,** 4119–4129.

Züst, B., and Dixon, K. E. (1975). The effect of u.v. irradiation of the vegetal pole of *Xenopus laevis* eggs on the presumptive primordial germ cells. *J. Embryol. Exp. Morphol.* **34,** 209–220.

Index

A

Aboral plate, sea urchin, 9
Actin microfilaments, endoplasmic reticulum clusters and, 136
Adenomatous polyposis coli protein, 14
Adhesion molecules, *see* Neural cell adhesion molecule
Adventive embryogenesis, 78–80
Alkaline phosphatase, 23
Allocation, 1
AML1 gene, 49
Amphibian germ plasm
   primordial germ cell formation, 164–165
   ultrastructure, 162
amp1 mutant, 77
Animal-vegetal axis patterning, sea urchin
   cell-cell interactions
      mesomere derivatives, 23–24
      micromere signaling, 18–21
      Notch pathway and mesoderm specification, 21–22
      vegetal-to-animal signaling cascade, 22–23
   maternal patterning
      β-catenin, 13–17, 33–35
      evidence of maternal polarity, 10–11
      micromere specification, 11–13
      model of, 32–33, 34–35
      VEB genes, 17–18
Apoptosis, germ cells and, 114
APX-1 protein, 171
Archenteron, sea urchin, 3
Arylsulfatase, 32
ATML1 gene, 69–70
Auxin, polar transport
   cotyledon formation, 75
   diffusion theory, 75–76
   embryonic polarity, 72
   procambium differentiation, 71
Axin, 14
Axis patterning, plant, 72
Axis patterning, sea urchin
   animal-vegetal axis
      cell-cell interactions, 18–24
      maternal patterning, 10–18, 32–35
   model of, 32–36
   oral-aboral axis, 24–32
   overview of, 2–3

B

bep RNAs, 7
bicoid RNA, 174
Blastomere isolation/recombination studies, sea urchin, 25–28
BMP, *see* Bone morphogenetic protein family
BMP1-like metalloprotease, 28–29
BMP2/4 signaling, in sea urchin axis patterning, 28–29, 35
Bone morphogenetic protein (BMP) family
   hematopoietic stem cell specification, 52–53
   Mix genes and, 50
   sea urchin axis patterning, 28–29, 35
   Vent genes and, 51
   ventral mesoderm patterning, 47–48, 53–54, 55
Boule protein, 173

C

Cadherins
   LvG-cadherin, 14, 15
   testicular development, 118–119
Caenorhabditis elegans
   germ plasm ultrastructure, 162–163
   P granule components, 169–172
   primordial germ cell development and migration, 159–160
Calcium-induced calcium release, 130, 135
Calcium-releasing channels, egg endoplasmic reticulum, 127–129

Calcium transients
   egg fertilization, 126–127, 130–131
      calcium-induced calcium release, 130
      capacitative calcium entry, 131, 139
      endoplasmic reticulum clusters and, 135–136, 138–139, 145–146
      endoplasmic reticulum dynamics and, 145–146
      endoplasmic reticulum stability during, 139–141
      single and multiple, 141–145
   immature oocytes, sperm-induced, 132
Callus, somatic embryogenesis, 80–82
Capacitative calcium entry, 131, 139
β-Catenin, sea urchin embryo patterning, 13–17, 33–35
CD4, sperm vector-DNA interactions, 92
Cell-cell interactions, sea urchin
   animal-vegetal axis patterning
      mesomere derivatives, 23–24
      micromere signaling, 18–21
      Notch pathway and mesoderm specification, 21–22
      vegetal-to-animal signaling cascade, 22–23
   oral-aboral axis patterning, 25–28
Cell migration, germ cells, 112–114
chordin mutant, 47
Chordin protein, 47
Chromatin, sperm, 93–94
Ciliary neurotrophic factor, 121
cis-Regulatory regions
   CyIII genes, 29–31
   Spec2a gene, 31–32
c-kit gene, 110–115
CLAVATA (CLV) gene, 73, 74
cloche gene, 53, 55
Collagen, type IV, 121
COLL1α RNA, 7, 24
ConcanavalinA, 12
Concatemers, 96–97
Cortical granule exocytosis, 132
Cotyledons, 71
   formation, 74–75
COUP-TF protein, 7
Cyclic ADP ribose, 128–129
cyclin B RNA, 167–168
CyIIa mRNA, 6
CyIII genes, 9
   cis-regulatory regions, 29–31
Cytochalasin D, 136
Cytostatic factor, 130, 142

D

DEAD-box helicases, 171, 174, 175
DEAD-South, 174, 175
Determination, defined, 1–2
2,4-Dichlorophenoxyacetic acid (2,4-D), 80, 81
Diffusion theory, 75–76
DiI labeling, 134
Direct development, 2
disheveled gene, 13
DNA
   organization in sperm, 93
   sperm vectors and
      binding, 90–91, 94–96
      integration, 92–94, 97–98
      internalization, 91–92, 96–97
      rhodamine-tagging, 91, 95
DNA-binding protein, sperm vector-DNA interactions, 92
DNA polymerase, exogenous DNA integration, 93
DNA repair enzymes, exogenous DNA integration, 97
Dorsal-vegetal signaling center, sea urchin, 13
Drosophila
   polar granules, 161
   pole cells
      development and migration, 159
      genes, 167–169
      pole plasm and, 163–164

E

Ectoderm, specification in sea urchin, 26–28
Ecto V protein, 26, 27–28
EGFII gene, 9
Egg activation, 126–127, 142
Eggs, *see also* Oocyte maturation
   activation, 126–127, 142
   calcium transients, 126–127, 130–131
      calcium-induced calcium release, 130
      capacitative calcium entry, 131, 139
      endoplasmic reticulum clusters and, 135–136, 138–139, 145–146
      endoplasmic reticulum dynamics and, 145–146
      endoplasmic reticulum stability during, 139–141
      single and multiple, 141–145
   endoplasmic reticulum

Index

calcium-releasing channels, 127–129
clusters, 134–136, 138–139, 145–146
future research directions, 146–147
pacemaker regions, 145, 146
polarity and, 134, 146–147
Embryo development, plant
  embryogenesis
    adventive, 78–79
    changing levels of embryogenic competence, 82–83
    growth and morphogenesis in, 62
    seed development and, 63–65
    somatic, 80–82
    zygotic, 78
  embryo-specific genes, 76–78
  histogenesis
    cell fate specification, 66–68
    maternal effects, 65–66
    overview of, 65
    procambium differentiation, 70, 71
    radial patterning, molecular mechanism, 68–70
  initial cell concept
    plant cell lineages, 68
    root meristems, 72–73
    shoot meristems, 73
  organogenesis
    cotyledon formation, 75
    diffusion theory and, 75–76
    embryonic polarity, 72
    root meristem, 72–73
    shoot meristem, 73–75
  overview of, 61–62, 83
Embryogenesis, plant
  adventive, 78–79
  changing levels of embryogenic competence, 82–83
  growth and morphogenesis in, 62
  seed development and, 63–65
  somatic, 80–82
  zygotic, 78
Embryogenic competence, plant, 82–83
Embryo patterning, sea urchin
  animal-vegetal axis patterning
    cell-cell interactions, 18–24
    maternal patterning, 10–18, 32–35
  model of, 32–36
  oral-aboral axis patterning, 24–32
  overview of, 2–3
EMF genes, 77–78
emf mutants, 77–78

Endoderm
  Mix genes, 50
  specification in sea urchin, 18–21, 22–23
Endo16 marker gene, 10, 18–19, 20, 23
Endoplasmic reticulum
  DiI labeling, 134
  egg
    calcium-releasing channels, 127–129
    calcium wave generation, 138–139
    clusters, 134–136, 138–139, 145–146
    future research directions, 146–147
    pacemaker regions, 145, 146
    polarity and, 134, 146–147
    stability during calcium transients, 139–141
  oocyte maturation
    IP_3 receptor distribution and, 136–138
    reorganization during, 133, 134–136
Estradiol, 121
ETS proteins, 12
Extracellular matrix, gonocyte development and, 121–122

F

Fate maps
  hematopoietic cell lineage, 48–49
  sea urchin, 3, 6–7
Fertilization, calcium transients, 126–127, 130–131
  calcium-induced calcium release, 130
  capacitative calcium entry, 131, 139
  endoplasmic reticulum clusters and, 135–136, 138–139, 145–146
  endoplasmic reticulum dynamics and, 145–146
  endoplasmic reticulum stability during, 139–141
  single and multiple, 141–145
Fertilization potential, 127
Fibroblast growth factor-2, 121
Fibulins, 121
FIDDLEHEAD gene, 69
Fluorescence recovery after photobleaching, *see* FRAP
Follicle-stimulating hormone, Sertoli cells and, 106
Founder cells
  defined, 1
  sea urchin endoderm and ectoderm, 6
FRAP (fluorescence recovery after photobleaching), endoplasmic reticulum dynamics, 140

G

Gametophytes, adventive embryogenesis, 79
gata-1 gene, 53
Gcl protein, 168, 175
Gene expression maps, sea urchin, 7–10
Gene transfer
  receptor-mediated, 92
  sperm-mediated
    concerns regarding, 98
    DNA binding, 90–91, 94–96
    DNA integration, 92–94, 97–98
    DNA internalization, 91–92, 96–97
    DNA markers and, 91, 94–96
    overview of, 89–90
    restriction enzyme-mediated integration, 97–98
    strategies in, 94–98
    transient gene expression, 95–96
germ cell-less (gcl) gene, 167, 168, 175
Germ cells, *see also* Primordial germ cells; Sertoli cell-gonocyte interactions
  determinants, 156 (*see also* Germ plasm)
  perinatal development, 104
  stem cell factor and, 110–111
Germinal granules, 160–163
germline helicase genes, 171
Germ plasm
  enclave hypothesis, 169, 175
  history, 157
  molecules localized to
    Drosophila genes, 167–169
    localized RNAs in *Xenopus*, 172–174
    overview of, 165–167
    P granule components in *C. elegans*, 169–172
    possible roles of, 174–176
  primordial germ cell formation, 156
    experimental evidence, 163–165
    germ plasm-localized molecules, 165–174
    overview of, 156, 174–176
  ultrastructure, 160–163
    amphibians, 162
    C. elegans, 162–163
    Drosophila, 161
    Xenopus, 157–158, 162, 164–165, 172–174
GLH proteins, 171–172, 175
Gonad coalescence, 159
Gonad formation, *Drosophila,* 159
Gonocytes, neonatal development
  extracellular matrix, 121–122
  growth factors, 120–121
  Sertoli cell-gonocyte interactions
    c-kit gene, 110–115
    coculture model, 106–110
    growth factors, 121
    NCAM-based adhesion, 115–116
    NCAM expression, 116–118
    overview of, 122
    thyroid hormone regulation of NCAM, 118–120
    in vivo, 107
goosecoid gene, 50
Granulofibrillar material, 162
Green fluorescent protein, sperm-mediated gene transfer and, 95, 96
Ground tissue, 65, 70
Growth, in plant embryo development, 62
Growth factors, gonocyte development and, 120–121
GSK3 protein, 13–14, 15, 16, 18
GSKS-binding protein, 14
Gut, sea urchin, specification, 21

H

He gene, 17, 18
Hematopoietic stem cell formation, 45–46
  fate maps, 48–49
  future research directions, 55
  mesoderm induction and patterning, 46–48, 53–54, 55
  overview of, 53–55
  specification of, 52–53
  ventralizing homeobox genes and, 49–51
Histogenesis, plant
  cell fate specification, 66–68
  maternal effects, 65–66
  overview of, 65
  procambium differentiation, 70
  radial patterning, molecular mechanism, 68–70
Homeobox genes, ventralizing, 49–51
HpArs gene, 32
Hp-ets RNA, 12–13
Hpoe antigen, 27
HpOtxL transcription factor, 32
Hypocotyl, 71
Hypophyseal cells, 72

I

Indirect development, 2–3
Initial cell concept

plant cell lineages and, 68
root meristems, 72–73
shoot meristems, 73
Inositol 1,3,4,5-tetrakisphosphate (IP$_4$), 131
Inositol 1,4,5-triphosphate (IP$_3$)
  calcium release in eggs, 128
  endoplasmic reticulum clustering in eggs, 135–136
  oocyte sensitivity to during maturation, 133
  production in eggs at fertilization, 129
Inositol triphosphate receptors (IP$_3$ receptors)
  in egg endoplasmic reticulum, 127, 128
  redistribution during oocyte maturation, 136–138
  subtypes in mammalian eggs, 129–130
Intracytoplasmic sperm injection, 95–96
Involution, sea urchin gastrulation, 3
IP$_3$, *see* Inositol 1,4,5-triphosphate
IP$_4$, *see* Inositol 1,3,4,5-tetrakisphosphate
IP$_3$ receptors, *see* Inositol triphosphate receptors

K

Kit protein, 110–115
KNAT1 gene, 70
KNOTTED 1 (KN-1) gene, 73, 74
kuzbanian, 22

L

LEAFY COTYLEDON 1 (LEC 1) gene, 77
LEAFY (LFY) gene, 74
Leukemia inhibitory factor, 121
Lithium chloride, 14, 15, 16–17
LvG-cadherin, 14, 15
LvNotch protein, 21–22

M

Macromeres, sea urchin
  autonomous endo/mesoderm specification, 19–20
  micromere signaling and, 16, 20–21
Major histocompatibility gene complex (MHC), sperm vector-DNA interactions, 92
Marginal zone, vertebrate
  mesoderm induction, 46–47
  primitive blood and, 49
Maternal effects, in plant histogenesis, 65–66
Maternal patterning, in sea urchin
  β-catenin and, 13–17, 33–35
  evidence of, 10–11
  micromere specification, 11–13
  model of, 32–33, 34–35
  VEB genes, 17–18
Maturation promoting factor, 130, 142
medea mutant, 65–66
Meiosis, in oogenesis, 126
Mesoderm
  specification in sea urchin, 18, 19–20, 21–22
  vertebrate
    induction, 46–47
    patterning, 47–48, 53–54, 55
Mesomere derivatives, sea urchin, 23–24
mex genes, 171
MEX proteins, 169, 171
Mice, primordial germ cells, 176
Microfilaments, endoplasmic reticulum clusters and, 136
Micromeres, sea urchin
  gene expression in descendants, 8–9
  signaling
    β-catenin and, 15, 16
    endo/mesoderm specification, 18–21
    Notch pathway and, 21–22
    Notch protein downregulation, 34
    vegetal-to-animal signaling cascade, 22–23
  specification, 11–13
Microtubules, endoplasmic reticulum clusters and, 136
Mitochondrial cloud, 162
Mitochondrial RNA, germ plasm and, 168, 175
Mix genes, 49–50
MONOPTEROS (MP) gene, 70, 71
monopteros (mp) mutant, 71, 72, 75
Morphogenesis, in plant embryo development, 62
Morphogenetic fields, adventive embryogenesis and, 79–80
mtlrRNA gene, 167, 168, 174
mtsrRNA gene, 168

N

nanos (nos) gene, 167–168, 169, 174, 175
Nanos protein, 168–169
NCAM, *see* Neural cell adhesion molecule
Neural cell adhesion molecule (NCAM), in Sertoli cell-gonocyte interactions
  cell-cell adhesion, 115–116
  developmental pattern of expression, 116–118
  thyroid hormone regulation and, 118–120

Nieuwkoop center, sea urchin, 13
NO APICAL MERISTEM (NAM) gene, 74
Noggin protein, 28, 29
Nonskeletogenic mesoderm, specification in sea urchin, 19, 21–22
nos-2 RNA, 172
Notch protein/pathway, sea urchin
  downregulation, micromere signaling and, 34
  mesoderm specification, 21–22
  oral-aboral polarity and, 6–7
  vegetal plate axis formation, 9–10
Nuage material, 162
Nuclear envelope, germ plasm and, 175
Nuclear matrix, exogenous DNA integration and, 93

O

Oocyte maturation
  endoplasmic reticulum arrangement and reorganization, 133, 134–136
  IP_3 receptor redistribution, 136–138
  IP_3 sensitivity, 133
  overview of, 126
  sperm-induced calcium release, 132
Oogenesis, 126
Oral-aboral axis patterning, sea urchin
  blastomere isolation/recombination studies, 25–28
  BMP2/4 signaling, 28–29
  β-catenin pathway, 16–17, 33–34
  cis-regulatory regions of differentially expressed genes, 29–32
  early expressions, 24–25
  model of, 33, 34–35
  vegetal plate, 6–7
Oral ectoderm, gene expression in sea urchin, 9
Organogenesis, plant, 71
orthopedia gene, 9
oskar RNA, 164, 167, 169
Otx transcription factor, 31
Ovary, adventive embryogenesis in plants, 78–80

P

P3A transcription factor, 31
P-cadherin, 119
Peritubular cells, 106, 121–122
P1 factor, 30
pgl-1 gene, 171

PGL-1 protein, 171
P granules, 163
  components, 169–172
  nuclear pores and, 175
Phosphatidylinositol 4,5-biphosphate (PIP_2), 128
Phospholipase C (PLC), 128, 129, 136
pickle mutant, 78
pie-1 gene, 171
PIE-1 protein, 169, 171
PINFORMED 1 (PIN 1) gene, 75
pinformed (pin 1) mutant, 75
PIP_2, *see* Phosphatidylinositol 4,5-biphosphate
Plants, *see* Embryo development, plant
Platelet-derived growth factor, 121
PLC, *see* Phospholipase C
PlHbox12 gene, 7–8, 25
PMCs, *see* Primary mesenchyme cells
Polar auxin transport, *see* Auxin, polar transport
polar granule component (pgc) RNA, 167, 168
Polar granules, 160, 161
Pole cells, *see also* Primordial germ cells
  development, 159
  genes, 167–169
  pole plasm and, 163–164
  history, 157
  migration, 159
Pole plasm, *see also* Germ plasm
  history, 157
  incorporation in pole cells, 159
  pole cell formation
    experimental evidence, 163–164
    genes, 167–169
Polysialic acid, 115
Polyspermy, electrical block, 127
pos-1 gene, 171, 174
POS-1 protein, 169, 171
Primary mesenchyme cells, sea urchin, 24
  gene expression in, 8
  specification, 11–13
Primordial germ cells
  development and migration
    C. elegans, 159–160
    Drosophila, 159
    Xenopus, 158–159
  formation, epigenetic, 156
  formation, germ plasm and
    experimental studies, 163–165
    germ plasm-localized molecules, 165–174
    overview of, 156, 174–176
  mouse, 176
  transcriptional repression, 175

Index

Procambium, 65
  differentiation, 70, 71
Protein phosphatase 2A, 14
Protoderm, 65, 69–70
Pseudopods, in germ cells, 112–113
pSV2-CAT, 94
Pumilio protein, 167–168

R

Radial patterning, in plant embryo development, 68–70
Rana, primordial germ cell formation, 164, 165
Receptor-mediated gene transfer, 92
Restriction enzyme-mediated integration, 97–98
Rhodamine-tagged DNA, 91, 95
RNA-binding proteins, primordial germ cells and, 169, 172, 175
Root, embryonic, 71
Root meristem, ontogeny and function, 72–73
Ryanodine, 128–129
Ryanodine receptors, in egg endoplasmic reticulum, 127, 128, 129

S

SCF, *see* Stem cell factor
scl gene, 52–53
SCL transcription factor, 52–53, 54–55
Scutellum, 75
Sea urchin
  axes of asymmetry, 2
  embryo patterning
    animal-vegetal axis, 10–24
    cell-cell interactions, 18–24
    compared to *Xenopus*, 35–36
    maternal patterning, 10–18, 32–35
    model of, 32–36
    oral-aboral axis, 24–32
    overview of, 2–3
  fate maps, 3, 6–7
  gene expression maps, 7–10
Secondary mesenchyme cells, sea urchin
  gene expression, 9
  specification, 19
Seed development, plant embryogenesis and, 63–65
Sertoli cell-gonocyte coculture model, 106–110
Sertoli cell-gonocyte interactions
  c-kit gene, 110–115
  coculture model, 106–110
  growth factors, 121

NCAM-based adhesion, 115–116
NCAM expression, 116–118
overview of, 122
thyroid hormone regulation of NCAM, 118–120
Sertoli cells
  differentiation, thyroid hormone and, 106, 117–118
  growth factors and, 121
  perinatal development, 104, 106
Sex-lethal (Sxl) gene, 167, 168
Shoot, embryonic, 71
Shoot meristem, ontogeny and function, 73–75
SHOOT MERISTEMLESS (STM) gene, 73, 76–77
short integument 1 (sin1) mutant, 65
siamois gene, 50
Skeletogenic mesenchyme, specification in sea urchin, 11–13
Smad5 protein, 47
SM50 gene, 8
Somatic embryogenesis, 80–82
somitobun gene, 47
spadetail gene, 46, 53
Spawning, egg calcium dynamics, 143
SpCOUP-TF RNA, 24
SpCOUP-TF transcription factor, 31
Spec genes, 9, 26
Specification, defined, 2
Spec1 protein, 27–28
Spemann organizer, 13
  in vertebrates, 49
  mesoderm induction, 46–47
Sperm, chromatin in, 93–94
Sperm-mediated gene transfer, *see* Gene transfer, sperm-mediated
SpEts4 gene, 18
SpEts4 protein, 18
SpGCF1 transcription factor, 30
SpHbox1 gene, 9
SpKrox-1 gene, 23
SpMyb transcription factor, 30, 31
SpOtx transcription factors, 31–32
SpP3A2 transcription factor, 30
SpSHR2 transcription factor, 31
Stem cell factor (SCF)
  germ cells and, 110–111
  Sertoli cell-gonocyte interactions and, 111–114
Stomodeum, 27
sus2 mutant, 67

Suspensor cells, 67–68
swirl gene, 47

T

TCG/LEF-1 protein, 14–15, 16
Testicular development
  neonatal, cell-cell interactions
    c-kit gene, 110–115
    extracellular matrix, 121–122
    growth factors, 120–121, 121
    NCAM-based adhesion, 115–116
    NCAM expression, 116–118
    overview of, 122
    Sertoli cell-gonocyte coculture model, 106–110
    thyroid hormone regulation of NCAM, 118–120
  perinatal
    germ cells, 104
    Sertoli cells, 104, 106
Thimerosal, 133
Thyroid hormone
  NCAM regulation and, 118–120
  Sertoli cell differentiation and, 106, 117–118
Tolloid-like metalloprotease, 28–29
Topoisomerase II, 93
Transgenic animals
  mosaicism, 94
  sperm-mediated gene transfer, 89–90
TransgenICSI, 95–96
Tudor protein, 167, 169
Twine protein, 173
twn mutants, 67

U

Ubiquitin/proteosome pathway, 14
UH2–95 antigen, 28

V

Vasa protein, 167, 169, 174, 175
VEB genes, 7, 8, 17–18
Vegetal cells, sea urchin, 3, 6
  oral-aboral axis patterning, 26–28
  vegetal-to-animal signaling cascade, 22–23, 34
Vegetal plate, sea urchin
  fate map, 6–7
  gene expression in, 9–10
VegT RNA, 174

VegT transcription factor, 46, 53
Vent genes, 51
Ventral blood island, 48, 49
Ventralizing homeobox genes, 49–51
Ventral mesoderm patterning, BMP pathway and, 47–48, 53–54, 55
Very early blastula genes, *see* VEB genes

W

Wnt/β-catenin pathway, sea urchin embryo patterning, 13–17, 33–35
WUSCHEL (WUS) gene, 74, 76–77

X

Xaml1 gene, 49
Xcat2 protein, 172–173
Xcat2 RNA, 172–173
Xdazl protein, 173, 174
Xdazl RNA, 173
Xenopus
  axis patterning, 35
  germ plasm, 157–158, 164–165
    localized RNAs, 172–174
    ultrastructure, 162
  hematopoietic stem cell formation
    fate maps, 48–49
    mesoderm induction and patterning, 46–48
    SCL transcription factor, 52–53
    ventralizing homeobox genes, 49–51
  primordial germ cells
    development and migration, 158–159
    germ plasm and, 164–165
Xklp-1 protein, 158
Xlsirts RNA, 173
Xpat RNA, 173–174
xtc mutants, 77
Xvent genes, 51
XWnt-8 protein, 15
Xwnt11 RNA, 173

Z

Zebrafish, hematopoietic stem cell formation, 46–48, 52
Zinc finger proteins, primordial germ cells and, 167, 169, 171
ZWILLE (ZLL) gene, 74, 76–77
Zygote, plant
  asymmetric division, 66–67
  embryogenesis, 78

Contents of Previous Volumes

Volume 45

1. **Development of the Leaf Epidermis**
   Philip W. Becraft

2. **Genes and Their Products in Sea Urchin Development**
   Giovanni Giudice

3. **The Organizer of the Gastrulating Mouse Embryo**
   Anne Camus and Patrick P. L. Tam

4. **Molecular Genetics of Gynoecium Development in *Arabidopsis***
   John L. Bowman, Stuart F. Baum, Yuval Eshed, Joanna Putterill, and John Alva

5. **Digging out Roots: Pattern Formation, Cell Division, and Morphogenesis in Plants**
   Ben Scheres and Renze Heidstra

Volume 46

1. **Maternal Cytoplasmic Factors for Generation of Unique Cleavage Patterns in Animal Embryos**
   Hiroki Nishida, Junji Morokuma, and Takahito Nishikata

2. **Multiple Endo-1,4-β-D-glucanase (Cellulase) Genes in *Arabidopsis***
   Elena del Campillo

3. **The Anterior Margin of the Mammalian Gastrula: Comparative and Phylogenetic Aspects of Its Role in Axis Formation and Head Induction**
   Christoph Viebahn

4. **The Other Side of the Embryo: An Appreciation of the Non-D Quadrants in Leech Embryos**
   David A. Weisblat, Françoise Z. Huang, Deborah E. Isaksen, Nai-Jia L. Liu, and Paul Chang

5 Sperm Nuclear Activation during Fertilization
Shirley J. Wright

6 Fibroblast Growth Factor Signaling Regulates Growth and Morphogenesis at Multiple Steps during Brain Development
Flora M. Vaccarino, Michael L. Schwartz, Rossana Raballo, Julianne Rhee, and Richard Lyn-Cook

Volume 47

1 Early Events of Somitogenesis in Higher Vertebrates: Allocation of Precursor Cells during Gastrulation and the Organization of a Meristic Pattern in the Paraxial Mesoderm
Patrick P. L. Tam, Devorah Goldman, Anne Camus, and Gary C. Shoenwolf

2 Retrospective Tracing of the Developmental Lineage of the Mouse Myotome
Sophie Eloy-Trinquet, Luc Mathis, and Jean-François Nicolas

3 Segmentation of the Paraxial Mesoderm and Vertebrate Somitogenesis
Olivier Pourquié

4 Segmentation: A View from the Border
Claudio D. Stern and Daniel Vasiliauskas

5 Genetic Regulation of Somite Formation
Alan Rawls, Jeanne Wilson-Rawls, and Eric N. Olsen

6 Hox Genes and the Global Patterning of the Somitic Mesoderm
Ann Campbell Burke

7 The Origin and Morphogenesis of Amphibian Somites
Ray Keller

8 Somitogenesis in Zebrafish
Scott A. Holley and Christiane Nüsslein-Volhard

9 Rostrocaudal Differences within the Somites Confer Segmental Pattern to Trunk Neural Crest Migration
Marianne Bronner-Fraser

Volume 48

1. **Evolution and Development of Distinct Cell Lineages Derived from Somites**
   Beate Brand-Saberi and Bodo Christ

2. **Duality of Molecular Signaling Involved in Vertebral Chondrogenesis**
   Anne-Hélène Monsoro-Burq and Nicole Le Douarin

3. **Sclerotome Induction and Differentiation**
   Jennifer L. Dockter

4. **Genetics of Muscle Determination and Development**
   Hans-Henning Arnold and Thomas Braun

5. **Multiple Tissue Interactions and Signal Transduction Pathways Control Somite Myogenesis**
   Anne-Gaëlle Borycki and Charles P. Emerson, Jr.

6. **The Birth of Muscle Progenitor Cells in the Mouse: Spatiotemporal Considerations**
   Shahragim Tajbakhsh and Margaret Buckingham

7. **Mouse–Chick Chimera: An Experimental System for Study of Somite Development**
   Josiane Fontaine-Pérus

8. **Transcriptional Regulation during Somitogenesis**
   Dennis Summerbell and Peter W. J. Rigby

9. **Determination and Morphogenesis in Myogenic Progenitor Cells: An Experimental Embryological Approach**
   Charles P. Ordahl, Brian A. Williams, and Wilfred Denetclaw

Volume 49

1. **The Centrosome and Parthenogenesis**
   Thomas Küntziger and Michel Bornens

2. **γ-Tubulin**
   Berl R. Oakley

3 γ-Tubulin Complexes and Their Role in Microtubule Nucleation
   Ruwanthi N. Gunawardane, Sofia B. Lizarraga, Christiane Wiese, Andrew Wilde, and Yixian Zheng

4 γ-Tubulin of Budding Yeast
   Jackie Vogel and Michael Snyder

5 The Spindle Pole Body of *Saccharomyces cerevisiae:* Architecture and Assembly of the Core Components
   Susan E. Francis and Trisha N. Davis

6 The Microtubule Organizing Centers of *Schizosaccharomyces pombe*
   Iain M. Hagan and Janni Petersen

7 Comparative Structural, Molecular, and Functional Aspects of the *Dictyostelium discoideum* Centrosome
   Ralph Gräf, Nicole Brusis, Christine Daunderer, Ursula Euteneuer, Andrea Hestermann, Manfred Schliwa, and Masahiro Ueda

8 Are There Nucleic Acids in the Centrosome?
   Wallace F. Marshall and Joel L. Rosenbaum

9 Basal Bodies and Centrioles: Their Function and Structure
   Andrea M. Preble, Thomas M. Giddings, Jr., and Susan K. Dutcher

10 Centriole Duplication and Maturation in Animal Cells
   B. M. H. Lange, A. J. Faragher, P. March, and K. Gull

11 Centrosome Replication in Somatic Cells: The Significance of the G_1 Phase
   Ron Balczon

12 The Coordination of Centrosome Reproduction with Nuclear Events during the Cell Cycle
   Greenfield Sluder and Edward H. Hinchcliffe

13 Regulating Centrosomes by Protein Phosphorylation
   Andrew M. Fry, Thibault Mayor, and Erich A. Nigg

14 The Role of the Centrosome in the Development of Malignant Tumors
   Wilma L. Lingle and Jeffrey L. Salisbury

15 The Centrosome-Associated Aurora/Ipl-like Kinase Family
   T. M. Goepfert and B. R. Brinkley

16 **Centrosome Reduction during Mammalian Spermiogenesis**
   G. Manandhar, C. Simerly, and G. Schatten

17 **The Centrosome of the Early *C. elegans* Embryo: Inheritance, Assembly, Replication, and Developmental Roles**
   Kevin F. O'Connell

18 **The Centrosome in *Drosophila* Oocyte Development**
   Timothy L. Megraw and Thomas C. Kaufman

19 **The Centrosome in Early *Drosophila* Embryogenesis**
   W. F. Rothwell and W. Sullivan

20 **Centrosome Maturation**
   Robert E. Palazzo, Jacalyn M. Vogel, Bradley J. Schnackenberg, Dawn R. Hull, and Xingyong Wu